Andrej Jentsch

A novel exergy-based concept of thermodynamic quality

AF099790

Andrej Jentsch

A novel exergy-based concept of thermodynamic quality

Development of a novel thermodynamic concept and its application to energy system evaluation and process analysis

Südwestdeutscher Verlag für Hochschulschriften

Impressum/Imprint (nur für Deutschland/ only for Germany)
Bibliografische Information der Deutschen Nationalbibliothek: Die Deutsche Nationalbibliothek verzeichnet diese Publikation in der Deutschen Nationalbibliografie; detaillierte bibliografische Daten sind im Internet über http://dnb.d-nb.de abrufbar.
 Alle in diesem Buch genannten Marken und Produktnamen unterliegen warenzeichen-, marken- oder patentrechtlichem Schutz bzw. sind Warenzeichen oder eingetragene Warenzeichen der jeweiligen Inhaber. Die Wiedergabe von Marken, Produktnamen, Gebrauchsnamen, Handelsnamen, Warenbezeichnungen u.s.w. in diesem Werk berechtigt auch ohne besondere Kennzeichnung nicht zu der Annahme, dass solche Namen im Sinne der Warenzeichen- und Markenschutzgesetzgebung als frei zu betrachten wären und daher von jedermann benutzt werden dürften.

Verlag: Südwestdeutscher Verlag für Hochschulschriften GmbH & Co. KG
Dudweiler Landstr. 99, 66123 Saarbrücken, Deutschland
Telefon +49 681 37 20 271-1, Telefax +49 681 37 20 271-0
Email: info@svh-verlag.de
Zugl.: Berlin, TU, Diss., 2010

Herstellung in Deutschland:
Schaltungsdienst Lange o.H.G., Berlin
Books on Demand GmbH, Norderstedt
Reha GmbH, Saarbrücken
Amazon Distribution GmbH, Leipzig
ISBN: 978-3-8381-1821-5

Imprint (only for USA, GB)
Bibliographic information published by the Deutsche Nationalbibliothek: The Deutsche Nationalbibliothek lists this publication in the Deutsche Nationalbibliografie; detailed bibliographic data are available in the Internet at http://dnb.d-nb.de.
 Any brand names and product names mentioned in this book are subject to trademark, brand or patent protection and are trademarks or registered trademarks of their respective holders. The use of brand names, product names, common names, trade names, product descriptions etc. even without a particular marking in this works is in no way to be construed to mean that such names may be regarded as unrestricted in respect of trademark and brand protection legislation and could thus be used by anyone.

Publisher: Südwestdeutscher Verlag für Hochschulschriften GmbH & Co. KG
Dudweiler Landstr. 99, 66123 Saarbrücken, Germany
Phone +49 681 37 20 271-1, Fax +49 681 37 20 271-0
Email: info@svh-verlag.de

Printed in the U.S.A.
Printed in the U.K. by (see last page)
ISBN: 978-3-8381-1821-5

Copyright © 2010 by the author and Südwestdeutscher Verlag für Hochschulschriften GmbH & Co. KG and licensors
All rights reserved. Saarbrücken 2010

Contents

Kurzzusammenfassung 8
Abstract 8

Introduction   **10**

1 State of the art   **11**
  1.1 What is Exergy? 11
    1.1.1 The reference state 11
    1.1.2 Different types of exergy 12
    1.1.3 Kinetic and Potential Exergy 13
    1.1.4 Physical exergy 13
    1.1.5 Chemical exergy 14
    1.1.6 Total exergy associated with mass transfers 15
    1.1.7 Exergy associated with energy transfers 15
    1.1.8 The exergy balance 16
  1.2 Applications of the exergy concept 17
    1.2.1 Developments in exergy analysis 18
  1.3 Thermodynamic energy system evaluation methods 19

2 The exergy-based transformability concept   **21**
  2.1 Introduction 21
  2.2 Splitting exergy into transformability and transformation energy for mass and energy flows 22
    2.2.1 Transformation energy 22
    2.2.2 The compensation heat flow 24
    2.2.3 Transformability 25
  2.3 Transformation energy and transformability associated with conductive heat flows . . 25
  2.4 Transformation energy and transformability associated with mass flows 29
    2.4.1 Thermal transformation energy and transformability associated with mass flows   29
    2.4.2 Mechanical transformation energy and transformability associated with mass flows 33
    2.4.3 Reactive transformation energy and transformability associated with mass flows   37
    2.4.4 Nonreactive transformation energy and transformability associated with mass flows 39
  2.5 Effective thermal transformability of differences in thermal transformation energy . . 43
  2.6 Average transformability and total transformation energy 45

3 Evaluation of energy supply systems and thermodynamic analysis using the transformability concept   **48**
  3.1 The transformation energy balance 48
  3.2 Transformability ratio and transformation energy efficiency 50

		3.2.1	Exergetic efficiency as a basis for transformability based evaluation ratios	50
		3.2.2	Definitions	51
		3.2.3	Considering compensation heat flows in transformation energy efficiency	54
		3.2.4	Interpretation of transformation energy efficiency and transformability ratio	57
	3.3	Setting evaluation boundaries for a cross-technology comparison of supply systems	59
		3.3.1	Specifying the supply target	59
		3.3.2	Defining cross-comparable subsystems	60
		3.3.3	Evaluating energy supply technologies	62
	3.4	Evaluation of the heat output from combined heat and power plants		62
		3.4.1	Derivation of the attribution of a fuel share to heat from CHP processes	63

4 Application of the transformability evaluation and analysis method — 66

- 4.1 Assessment of energy supply systems — 66
- 4.2 Results of comparative transformability assessment for examples from heating and cooling — 67
  - 4.2.1 Heating systems — 67
  - 4.2.2 Cooling systems — 70
- 4.3 Influence of reference temperature on the evaluation of thermal supply systems — 72
- 4.4 Effects of heat losses and pressure drops on the evaluation of some basic processes — 78
  - 4.4.1 Heat Exchanger — 81
  - 4.4.2 Boiler — 81
  - 4.4.3 Compression heat pump — 81
  - 4.4.4 Compression refrigeration machines — 82
  - 4.4.5 Heat engine — 82
  - 4.4.6 Expander — 83
  - 4.4.7 Compressor — 83
  - 4.4.8 Summary — 84
- 4.5 Example - Analysis of a vapor-compression cascade refrigeration machine — 84
- 4.6 The ExergyFingerprint - a transformability-based graphical assessment tool — 88
- 4.7 Defining the term "LowEx" by using the transformability concept — 91

5 Discussion and Conclusion — 94

- 5.1 Transformation energy and transformability — 94
- 5.2 Transformability assessment and analysis — 95
  - 5.2.1 The transformation energy balance and the compensation heat flow — 95
  - 5.2.2 Interpretation of the results of transformability evaluation and analysis — 96
- 5.3 Advantages and Disadvantages of the transformability concept and the transformability evaluation and analysis method — 99
  - 5.3.1 Advantages of the transformability concept — 99
  - 5.3.2 Advantages of transformability assessment and analysis — 100
  - 5.3.3 Disadvantages — 102
- Outlook — 103

Contents

Summary   105

Zusammenfassung   108

Nomenclature   111
    Greek letters 111
    Superscripts 112
    Subscripts 112
    Abbreviations 114

List of Figures   116

List of Tables   118

Bibliography   119

Appendices   123
    A1   Summary of the exergy interpretation underlying this work 123
         A1.1   Avoiding misconceptions 123
    A2   Calculation of chemical exergy 124
    A3   On the use of separate types of exergy for exergy analysis 126
    A4   Summary of transformability and transformation energy definitions for practical application 130
    A5   Summary of expressions for effective thermal transformabilities 132
    A6   Calculation of ideally required heat transfers 133
         A6.1   Evaluation of refrigeration machines 133
         A6.2   Evaluation of heat exchangers 133
         A6.3   Evaluation of heat engines 134
         A6.4   Evaluation of compressors and expanders 134
    A7   Calculation basics for the evaluation of heat production from combined heat and power processes 137
         A7.1   Calculation of the average transformability associated with heat from CHP delivered by district heating 139
    A8   Calculating evaluation ratios for the evaluation of heat and cold supply systems . . . 141
    A9   Summary of the transformability assessment method for energy supply technologies . 142
    A10 Calculations for comparative evaluation of supply systems for domestic heating 143
         A10.1 Boiler 144
         A10.2 Heat from a geothermal heat source 146
         A10.3 Ground-source heat pump 147
         A10.4 Block Combined Heat and Power plant 149
    A11 Calculations for comparative evaluation of supply systems for domestic cooling 152
         A11.1 Compression refrigeration machine 153
         A11.2 Direct seawater cooling 156
         A11.3 Absorption cooling using waste heat 158

Contents

A12	On the use of average reference temperature	162
A13	Equations for transformability analysis of some common components	164
A14	Analysis of a vapor-compression cascade refrigeration machine	165
A15	Basic data for the calculation of the exemplary ExergyFingerprints	173

Contents

Kurzzusammenfassung

In dieser Arbeit wird ein neues Konzept entwickelt, welches es ermöglicht, die Exergie von Masse- und Energieströmen als Produkt aus thermodynamischer Qualität (Wandelbarkeit) und Quantität (Wandlungsenergie) zu interpretieren. Aufbauend auf dieser Aufspaltung der Exergie in zwei neue Größen wird eine Bewertungsmethode vorgeschlagen, welche es möglich machen soll verschiedene Versorgungstechnologien auf Basis der Exergie transparenter als bisher zu vergleichen. Dabei werden die Wandlungsenergieeffizienz , welche als Grad der externen Güte interpretierbar ist und das Wandelbarkeitsverhältnis, welches sich als Grad der Prozesseignung verstehen lässt, verwendet. Das Produkt der beiden neuen Größen ist die exergetische Effizienz. Zusätzlich wird ein strukturiertes Vorgehen für die Definition der Bilanzgrenzen von Versorgungssystemen vorgeschlagen. Die Besonderheiten der Kraft-Wärme-Kopplung sowie nicht-speicherbarer erneuerbarer Energien werden dabei berücksichtigt. Die Bewertungsmethode wird beispielhaft auf Wärme- und Kälteversorgungssysteme angewendet. Weiterhin wird die Eignung der neuen Methode zur thermodynamischen Analyse anhand von einfachen thermodynamischen Prozessen sowie einer Dampf-Kompressionskältemaschine untersucht. Die Dissertation wird mit einer Diskussion der Vor- und Nachteile der neuen Methode im Vergleich zu ausschließlich exergetischer Bewertung und Analyse abgeschlossen.

Abstract

In this work a novel concept is developed that allows to interpret exergy associated with mass or energy transfers as a product of thermodynamic quality (transformability) and quantity (transformation energy). Based on this splitting of exergy into two novel properties an evaluation method is suggested that allows a transparent exergy-based comparison of different energy supply technologies using transformation energy efficiency, which can be interpreted as a a measure of external sophistication and transformability ratio, which indicates process suitability. The product of the two novel evaluation ratios is exergetic efficiency. Additionally, a consistent structured procedure for the evaluation of energy supply systems for domestic heating and cooling is laid out that includes a comprehensive rule-based boundary definition and an exergy-based attribution of fuel to heat from combined heat and power processes. The developed method is exemplary applied to supply systems for domestic heating and cooling as well as used for the analysis of some basic thermodynamic processes and a vapor-compression cascade refrigeration machine. The dissertation concludes with a discussion of the advantages and disadvantages of the novel analysis and evaluation method in comparison to an exclusively exergetic evaluation and analysis.

Clarification - On the use of the word "exergy" in this work

The term "exergy" is commonly used as a characteristic that can be associated with closed systems and with flows. This work focuses on system assessment based on exergy associated with flows, which are considered at constant parameters. Since exergy associated with flows is calculated differently from exergy associated with closed systems, the term "exergy" will be used only for exergy associated with flows in this dissertation.

Introduction

How to assess thermodynamic quality?

Efficiency is a widely discussed topic as fossil ressources become less available and emissions from the energy supply sector are increasingly considered problematic for the global climate. Usually, the discussion of efficiency is based on the energy concept. The first law of thermodynamics states that energy cannot be destroyed. Consequently, energy can only be converted from one form to another and only losses through the considered system boundaries influence energy efficiency. Experience shows that not all energy forms can be fully converted into other energy forms. Based on this experience the second law of thermodynamics has been developed. It states according to Lord Kelvin (Dunning-Davies, 1965) that it is impossible to convert heat completely into work in a cyclic process in the absence of other effects. As a consequence, in thermodynamic theory some forms of energy are considered to be more useful than others. Thus, aside from the quantitative aspects that can be balanced using the first law of thermodynamics, energy transfers can also be associated with a quality aspect. This image that energy also has a quality aspect, is easy to understand as it summarizes in a simply communicable way the fact that not all forms of energy are equally interesting for technological applications. Up to date only the exergy concept provides a scientifically consistent, process-independent and universal approach that takes into account the quality aspect of energy. Exergy associated with energy transfers is a measure of the theoretical ability to generate work using a considered flow if allowing interaction only with a predefined environment. However, the exergy associated with such an energy transfer is no measure of its quality alone. Exergy always incorporates the quantitative aspect of the transfer as well, thus making it difficult to separate the influence of the "size" of the flow from its "quality".

For many applications the use of such a combined property provides significant improvements over the assessment of energy alone. For example, the usefulness of compressed ideal gas flows at ambient temperature cannot be described using energy, but it can be associated with an exergy value. Such a mechanical exergy flow is in principle comparable with an exergy flow associated with a heat transfer. The universal applicability and the consistent incorporation of second law effects are the essential benefits of exergy over energy. Still, the merging of quantitative and qualitative aspects within one property also results in a loss of information. A thermodynamic loss within a process can be caused by a loss flow through a system boundary or by internal exergy destruction. Energy analysis on the other hand can only indicate losses with respect to the system boundaries.

In this work, a consistent and universal theory of exergy-based measures for thermodynamic quantity and quality is developed. It can be a basis for a scientifically sound answer to the search for a measure of "thermodynamic quality". Based on a set of new properties that allow separate association of considered transfers with an energetic quantity and a thermodynamic quality a novel exergy-based assessment and analysis method is developed that can provide a new perspective on thermodynamic performance.

1 State of the art

This chapter covers briefly the basics of exergy analysis, recent developments in exergy theory and the use of exergy as a means for thermodynamic assessment of energy systems. A short summary of the exergy interpretation that lays the basis for the theories developed in this dissertation can be found in appendix A1 on page 123.

1.1 What is Exergy?

The comparative evaluation of energy systems can be undertaken from various points of view. In engineering the thermodynamic performance is the technical criterion of central interest. One of the means to assess the thermodynamic performance of technical systems is exergy assessment and analysis. Exergy analysis „uses the conservation of mass and conservation of energy principles together with the second law of thermodynamics" (Moran and Shapiro, 2007, p. 329) for the analysis and assessment of technical systems, thus incorporating the two central thermodynamic principles within one property.

The word „exergy" has been introduced by Rant (1956) and stems from the Greek words *ex* (meaning "from") and *ergon* (meaning "work"). Exergy is also known under the names available energy, availability, essergy (Szargut et al., 1988) and *technische Arbeitsfähigkeit* which is German meaning „technical ability to do work" (Bosnjakovitch, 1935; Grassmann, 1951). The major difference of exergy to energy is, that it can be destroyed. Various slightly different definitions of exergy can be found in literature (Bosnjakovitch, 1935; Ahrendts, 1977; Szargut et al., 1988; Bejan et al., 1996; Dincer and Cengel, 2001; Moran and Shapiro, 2007, p.114). The most recent definition provided by Tsatsaronis (2007, p.249) states:

> „Exergy of a thermodynamic system (or stream) is the maximum theoretical useful work (shaft work or electrical work) obtainable as the system (or stream) is brought into complete thermodynamic equilibrium with the thermodynamic environment while the system (or stream) interacts with this environment only."[1]

1.1.1 The reference state

The definition of a reference environment is a premise for exergy calculation. Since the real environment is not totally in thermodynamic equilibrium only common components in often encountered concentrations and aggregate states can be used for the definition of the reference state (Szargut et al., 1988). Various approaches to define a reference state have been taken, especially regarding its chemical composition. However, all these approaches share the assumption of a homogeneous reference state (Wall, 1977). Furthermore, the reference environment is assumed to display reservoir characteristics, meaning that its intensive properties remain constant if interacting with a finite system

---

[1] The addition: (or stream) relates to mass and energy transfers (G. Tsatsaronis 2010, personal communication, 22 February)

1 State of the art

of interest (Wall, 1977). The natural environment is not in thermodynamic equilibrium. According to Ahrendts (1977) the specific exergy of oxygen would be more than thirty times higher if a complete equilibrium with the earths crust down to $1000\,m$ would be assumed. Since exergy is a property of the combined system, it can only be used as a quasi-property associated with the finite system or flow of interest, if the reference state, for which exergy values are compared, is the same for all systems or flows. In general the reference state can be arbitrarily chosen. However, for operations on earth's surface it is most sensible that the reference state reflects the state of the natural environment as closely as possible.

1.1.2 Different types of exergy

The fundamental differentiation of exergy is that between exergy associated with closed systems and exergy associated with energy or mass transfers. Since this work is only based on the application of the exergy concept to the evaluation of mass and energy transfers, the discussion will be focused on this type of exergy, while exergy associated with closed systems will not be part of the discussion.

Exergy associated with mass flows can be subdivided into different types of exergy which are functions of the deviation of a single intensive property of the flow from reference conditions. Slightly different terminologies have been used for the classification of the types of exergy (Fratzscher et al., 1986; Szargut et al., 1988). Following a recently published terminology by Tsatsaronis (2007, p. 249) the different types of exergy can be termed as:

1. Kinetic exergy - originating in a difference of velocity between the considered mass flow and the environment.

2. Potential exergy - which is associated with a potential of the considered mass flows in a force field (like the gravitational field of the earth) measured in relation to the potential of the environment.

3. Physical exergy

    a) Thermal exergy - which is a function of the difference in temperature between the flow under consideration and the environment. The thermal exergy of a flow at T, p should be calculated along the isobaric line at p - from state $[T, p]$ to state $[T_0, p]$

    b) Mechanical exergy - originating in a difference in pressure between the considered transfer and the environment. It should be calculated for a given state at T, p along the isothermal line at T_0 - from state $[T_0, p]$ to the reference state $[T_0, p_0]$ after thermal exergy has been evaluated.

4. Chemical exergy - which is evaluated at reference pressure and temperature.

    a) Reactive exergy - which originates from the capacity of a considered substance that is not part of the environment to react with components of the environment.

    b) Nonreactive exergy - that is associated with substances that are present in the reference environment but differ from it in concentration.

1 State of the art

5. Nuclear exergy - which is usually neglected for conventional engineering considerations and is only added here following Fratzscher et al. (1986) to complete the enumeration. It can be separated into the following two types:

   a) Fission exergy - Exergy associated with a nuclear fission reaction.

   b) Fusion exergy - Exergy associated with a nuclear fusion reaction.

Adding to these types of exergy that can be defined only for mass flows, mass-free energy transfers can also be associated with exergy, the most important types of exergy being:

- Exergy associated with conductive heat transfers
- Exergy associated with radiation
- Exergy associated with mechanical work

1.1.3 Kinetic and Potential Exergy

These types of exergy are associated with ordered movement or macroscopic elevation of particles of matter. The equations for kinetic and potential exergy equal those for potential and kinetic energy. These forms of exergy consequently do not contain an entropic term and can thus be handled easily. Because these forms of exergy are dependent only on the total mass of a flow and not a function of the specific substance or mixture, like all other types of exergy that are associated with mass flows, they can be simply added to any exergy equation.

Thus, in order to keep exergy equations compact and in order to focus on more complicated types of exergy, these types of exergy are set to zero for most considerations. The kinetic exergy E^{KN} can be calculated as a function of the considered mass m and its velocity c (Moran and Shapiro, 2000) :

$$E^{KN} = m \cdot e^{KN} = m \cdot \frac{c^2}{2}$$

Potential exergy E^{PT} is calculated as a function of mass, the gravitational acceleration g and the altitude z :

$$E^{PT} = m \cdot e^{PT} = m \cdot g \cdot z$$

The exergy of charged particles in an electric field that is considered potential exergy by Ignatenko et al. (2007) is not relevant for the thermodynamic considerations in this work and is therefore not discussed.

1.1.4 Physical exergy

Physical exergy is the sum of mechanical and thermal exergy. For an accurate exergy evaluation it is necessary to evaluate thermal and mechanical exergy separately (Lazzareto and Tsatsaronis,

1 State of the art

2006). The thermal exergy flow $\dot{E}^T$ associated with a mass flow of a pure substance or a mixture at temperature T and pressure p can be expressed as a function of the considered mass flow $\dot{m}$, the specific enthalpy h, the specific entropy s and the reference temperature T_0:

$$\dot{E}^T = \dot{m} \cdot e^T = \dot{m} \cdot \{h(T,p) - h(T_0,p) - T_0 \cdot [s(T,p) - s(T_0,p)]\} \tag{1.1}$$

Similarly, a mechanical exergy flow $\dot{E}^M$ can be expressed as a function of the abovementioned properties and the reference pressure p_0:

$$\dot{E}^M = \dot{m} \cdot e^M = \dot{m} \cdot \{h(T_0,p) - h(T_0,p_0) - T_0 \cdot [s(T_0,p) - s \cdot (T_0,p_0)]\} \tag{1.2}$$

If the mass flow is a mixture of different substances, thermal and mechanical exergy are generally calculated using specific enthalpy and entropy values of the mixture. For mass flows above reference pressure, thermal and mechanical exergy flows can be considered together as physical exergy flows $\dot{E}^{PH}$:

$$\dot{E}^{PH} = \dot{m} \cdot e^{PH} = \dot{m} \cdot \{h(T,p) - h(T_0,p_0) - T_0 \cdot [s(T,p) - s \cdot (T_0,p_0)]\} \tag{1.3}$$

Szargut (2005, p. 20) states that the mechanical exergy of mass flows below reference pressure is negative. As a consequence, physical exergy of mass flows below reference pressure can consist of thermal and mechanical exergy with different algebraic signs. It can thus be argued that a separate evaluation of the different types of exergy is often necessary to obtain sensible results as the use of physical exergy would otherwise lead to a factual subtraction of mechanical from thermal exergy. Appendix A3 on page 126 explains the necessity of a separate evaluation of the different types of exergy as a basis for a consistent and universal applicability of the transformability assessment method, which is developed in this work.

1.1.5 Chemical exergy

Chemical exergy summarizes exergy that is associated with mass flows at reference temperature and pressure due to differences in molecular structure and to differences in concentration. Moran and Shapiro (2000) give the following definition:

> „The chemical exergy is the maximum theoretical work that could be developed by a fuel cell into which a substance of interest enters at reference temperature T_0 and reference pressure p_0 and reacts completely with environmental components to produce environmental components."

A chemical exergy flow $\dot{E}^{CH}$ associated with a mass flow can be expressed as a sum of a reactive exergy flow $\dot{E}^R$ and a nonreactive exergy flow $\dot{E}^N$ (Bejan et al., 1996; Tsatsaronis, 2007).

$$\dot{E}^{CH} = \dot{E}^R + \dot{E}^N$$

Thus, if considering chemical exergy of a fuel not present in the environment as being the work generated by a reaction in a reversible fuel cell, the fuel enters the fuel cell as a pure substance at

1 State of the art

reference temperature and pressure. Molecules that are part of the reference environment can only be associated with nonreactive exergy due to concentration differences (Lukas, 2004), which can be changed by mixture and separation processes. Such substances would enter the ideal fuel cell at starting concentration and reference temperature and pressure and exit it at the concentration of the substance in the reference environment. The calculation of chemical exergy is presented in appendix A2 on page 124 ff. .

1.1.6 Total exergy associated with mass transfers

Total exergy is the sum of all types of exergy that are associated with a considered mass flow. A general equation for total exergy associated with a given mass flow could thus be expressed as:

$$\dot{E}^{TO} = \dot{m} \cdot \left[h - h_0 - T_0 \cdot (s - s_0) + \frac{c^2}{2} + gz + e^{CH} \right] \quad (1.4)$$
$$= \dot{m} \cdot \left(e^{PH} + e^{KN} + e^{PT} + e^{CH} \right)$$

If more than one substance is present in the considered mass flow the specific enthalpy and entropy and specific chemical exergy have to be considered for the mixture as a whole. Assuming an ideal mixture, the molar specific chemical exergy associated with the mixture flow can be calculated as function of the mole fractions x and the specific molar chemical exergy $\bar{e}^{CH}$ of the different substances j (Bejan et al., 1996):

$$\bar{e}^{CH} = \sum_j x_j \cdot \bar{e}_j^{CH} + \bar{R} \cdot T_0 \cdot \sum x_j \ln x_j$$

Usually, the expression of total exergy associated with a mass flow from Equation 1.4 is considered to be unboundedly valid. However, some types of exergy can in principle have values below zero, e.g. mechanical exergy for mass flows below reference pressure. Therefore, the summation of the different types of exergy could result in a factual subtraction of specific types of exergy (thermal, mechanical) from each other. Although for many cases the evaluation of mass flows using total exergy is unproblematic, in general the use of total exergy can lead to results that e.g. do not allow a sensible definition of exergetic efficiency for a heat pump. This is demonstrated in appendix A3 on page 126 ff. Consequently, in this thesis all forms of exergy will be considered separately.

1.1.7 Exergy associated with energy transfers

The most significant energy transfers that are not associated with mass flows are heat transfers, work transfers and energy transfer by thermal radiation. The exergy flow $\dot{E}^Q$ associated with a heat transfer at a constant temperature is defined as a function of the conductive heat transfer $\dot{Q}$ at its

1 State of the art

temperature T (Bosnjakovic and Knoche, 1998) as:

$$\dot{E}^Q = \left(1 - \frac{T_0}{T}\right) \cdot \dot{Q}$$

Since exergy is a measure of the amount of work ideally obtainable from a combined system of flow and environment, the exergy flow associated with a work flow $\dot{E}^W$ can be defined as:

$$\dot{E}^W = \dot{W}$$

Exergy associated with thermal radiation is discussed extensively by Bosnjakovic and Knoche (1998) and Petela (2003). However, the development of novel exergy-based properties requires a full understanding of the derivation of the different types of exergy. Without additional knowledge in radiation and photon physics, which are not part of engineering curricula, a profound understanding of exergy associated with thermal radiation seems very difficult. Since for most engineering applications exergy associated with radiation plays little or no role an extensive review of this type of exergy would exceed the scope of this thesis. Consequently, exergy associated with radiation is exempted from the following discussion and left for future investigation.

1.1.8 The exergy balance

The exergy balance is the basis for exergetic evaluation. For steady-state processes and systems it can be expressed as a function of the rate of exergy destruction $\dot{E}_D$, the sum of all exergy flows entering the system boundary $\sum \dot{E}_i$ and the sum of all exergy flows exiting the system boundary $\sum \dot{E}_e$ (Fratzscher et al., 1986):

$$\dot{E}_D = \sum \dot{E}_i - \sum \dot{E}_e$$

Using the fuel and product concept for exergetic efficiency definition, which has been presented first by Tsatsaronis (1984) and has been discussed more thoroughly by Tsatsaronis and Winhold (1985) and Bejan et al. (1996), the exergy balance can be expressed as a function of the exergy destruction $\dot{E}_D$, the sum of all of product exergy flows $\sum \dot{E}_P$, the sum of all exergy losses $\sum \dot{E}_L$ and the fuel exergy flow $\dot{E}_F$:

$$\dot{E}_F = \dot{E}_P + \dot{E}_L + \dot{E}_D$$

In this equation, the fuel term consists of the sum of exergy decreases and exergy inputs into the considered system. Bejan et al. (1996) state additionally that at a component level exergy increases that are not in accordance with the purpose of the component have to be subtracted from the fuel exergy term. However, a short discussion in subchapter 3.2.2 on page 51 explains why an inclusion of a subtraction into the fuel term is not sensible for the exergy-based concept developed in this dissertation. Instead, exergy increases that are not in direct accordance with the purpose of a considered component can be taken into account in the product term if they are a part of the fuel of another system component. If exergy increases that are not in accordance with the purpose of the component are not used, they can be interpreted as losses, which are considered neither in the fuel term nor in the product term.

Product exergy is defined as the sum of useful exergy outputs from the process and of exergy increases caused by the process.

Exergy satisfies the law of conservation only if reversible processes are considered.

1.2 Applications of the exergy concept

The exergy concept is used for a broad variety of applications. It is used for ecological modelling (Jorgensen, 1999; Susani et al., 2005), for the assessment of the technological aspects of sustainability (Wulf et al., 2000; Berthiaume et al., 2001; Rosen, 2002; Balocco et al., 2003; Lems et al., 2003; Hau, 2005; Rosen, 2008b) and for thermoeconomic analysis (Szargut et al., 1988; Bejan et al., 1996; Tsatsaronis and Park, 2002; Hebecker et al., 2004), which is the most established of the not purely thermodynamic applications of the exergy concept. Recently, the exergy concept has been included into an exergoenvironmental analysis (Meyer et al., 2009), which according to the authors reveals the extent to which each component of an energy conversion system is responsible for the overall environmental impact and allows to identify the sources of the impact[2].

However, the original area of application of the exergy concept is thermodynamic analysis and system evaluation (Tsatsaronis, 1999, p.93).

„Exergy analysis identifies the location, the magnitude and the sources of thermodynamic inefficiencies in a system."

Using the exergy concept, energy engineering tasks can be formulated under consideration of environmental conditions, but independent of the systems that are used or can be used to solve the considered technical problem. Exergy is well suitable for a thermodynamically just comparison of a variety of technologies (Franke, 1998).

Different exergy-based ratios can be applied to evaluate the performance of a technical system. The most important one is exergetic efficiency, which has been defined by Tsatsaronis (1984) as the ratio of the exergy flow $\dot{E}$ associated with the product (subscript P) to the exergy flow associated with the fuel (subscript F) of the process:

$$\varepsilon = \frac{\dot{E}_P}{\dot{E}_F} = 1 - \frac{\dot{E}_L + \dot{E}_D}{\dot{E}_F}$$

This definition requires a sensible definition of fuel and product as discussed in subsection 1.1.8 on the previous page.

Various other ratios have been defined in order to characterize a thermodynamic system (Fratzscher et al., 1986; Tsatsaronis, 1999, p.93). Based on an evaluation of six common ratios for the analysis of technical components Tsatsaronis (1999) comes to the conclusion that of the ones investigated

---
[2]None of these applications have been critically reviewed, since they are not relevant in the context of this work. The enumeration is solely intended to illustrate the variety of applications of the exergy concept.

1 State of the art

the only variable that unambiguously characterizes the performance of a component from a thermodynamic point of view is exergetic efficiency. Therefore, in this work only exergetic efficiency will be considered as a universally applicable assessment parameter that allows to characterize the thermodynamic performance of a component or system and to compare it with the performance of other similar components.

However, for the comparison of dissimilar components within a process the exergy destruction ratio y_D, which is defined as the ratio of the exergy destruction within the component j to the fuel input into the system:

$$y_D = \frac{\dot{E}_{D,j}}{\dot{E}_F}$$

is the most appropriate variable of the ones considered by Tsatsaronis (1999). In this case, the component boundary is defined in such a way that exergy is only destroyed within but not lost from the component.

1.2.1 Developments in exergy analysis

Based on the exergy concept, advanced assessment parameters have been developed that allow a more precise analysis of thermodynamic systems. These developments indicate that the theoretical development of the exergy concept is not yet finished.

An advanced approach to exergy analysis has been presented by Tsatsaronis and Park (2002). The authors develop a method for the distinction of avoidable and unavoidable exergy destruction and accordingly avoidable and unavoidable costs. The unavoidable exergy destruction is defined as the part of exergy destruction that remains present even if infinite investment costs for the considered components would be allowed. The unavoidable investment costs are defined as the lowest investment costs possible, even if this would result in the use of very inefficient versions of the relevant components. The avoidable exergy destruction and investment costs can be calculated by subtracting the unavoidable parts from the total values. The method presented by the authors allows an analysis of a given process with respect to its realistic improvement potential.

Another novel yet apparently not fully consistent concept for process analysis, which is claimed to be suitable especially for heat and matter exchange technologies, has been proposed by Chang and Chuang (2003). They define an extrinsic exergy loss as part of the total exergy loss due to deviation of the process from reversibility and an intrinsic exergy loss that describes the exergy loss of a reversible process due to deviation of the considered process from the assumption of perfect equilibrium within mass exchange processes. However, the authors do not differentiate between recoverable exergy loss and exergy destruction. Additionally, they do not discuss the fact that some components cannot be operated reversibly, such as a co-current heat exchanger with two different input temperatures or mixing chambers. It is therefore questionable whether the approach demonstrated by the authors is consistently usable based on the partially imprecise terminology applied.

Hebecker et al. (2004) presented a method for a hierarchically structured approach to exergy analysis. It is based on the determination of additive loss coefficients which essentially rate the exergy destruction and losses to the exergy input into a component. Using a significance factor which relates the exergy input into the component to the exergy input into the hierarchically superimposed system, a loss component can be calculated as a product of the significance factor and the loss coefficient. The hierarchically structured approach to exergy analysis appears to be a promising way of obtaining a meaningful set of exergy-based ratios suitable for the identification of problematic and less problematic components within complex hierarchical systems.

Recently, Morosuk and Tsatsaronis (2008) included the principle of avoidable and unavoidable exergy destruction (Tsatsaronis and Park, 2002) into a so called advanced exergy analysis. Additionally, they introduce a separation of exergy destruction into endogenous and exogenous exergy destruction. The endogenous exergy destruction is the part of exergy destruction caused only within the considered component if the rest of the considered process is considered to be reversible. The difference between the actual exergy destruction in the component and the endogenous exergy destruction is then termed exogenous exergy destruction. With the provided set of new exergy destruction sub-types (avoidable/unavoidable and endogenous/exogenous), an exergetic analysis can be performed with a higher precision thus providing results of significantly higher practical value. Recently Kelly et al. (2009) have compared different theoretical methods for the definition of exogenous and endogenous exergy destruction, thus further developing the advanced exergy analysis.

In addition to the developments in exergy analysis, first attempts at developing exergy-based properties have been made. Several authors (Nieuwlaar and Dijk, 1993; Bittrich and Hebecker, 1999; Petela, 2003; Utlu and Hepbasli, 2007; Rosen, 2008a; Xia et al., 2008) have mentioned a definition of an exergy-based quality measure. These first definitions have in common that they use a quality indicator that is essentially given as the ratio of the exergy flow to the enthalpy, heat or work flow it is associated with. Although this ratio can be useful in the evaluation of some flow types, such as conductive heat flows above reference temperature, it yields nonsense results for many other types of exergy. For example, using the exergy rate to energy rate ratio as a means to assess thermodynamic quality could lead to the misconception that a mechanical exergy flow, which can easily have an exergy rate to enthalpy[3] rate ratio exceeding ten, is many times more valuable than a work flow with an exergy rate to energy rate ratio of one. Furthermore, heat flows below reference temperature, for which the exergy rate to energy rate ratio is always below zero, have a negative and thus totally different quality than those above reference temperature for which this ratio has always positive values. Additionally, the absolute value of the exergy rate to energy rate ratio for heat flows below reference temperature can exceed that of work flows. Thus, it has to be concluded that the exergy rate to energy rate ratio cannot be considered a universally valid definition of thermodynamic quality.

1.3 Thermodynamic energy system evaluation methods

Energy system evaluation has the goal to identify optimization potential and can be used to compare processes and systems in respect to a selected dimension thus laying the basis for rational selection of the best technologies. Energy systems can be evaluated on a large scale of different aspects, such as economical, environmental, technical or social dimensions. The technically most relevant

---

[3]The reference state for the evaluation of enthalpy is assumed to be identical with the reference state assumed for the evaluation of exergy.

is thermodynamic performance. Basically energy systems deviate from a thermodynamic theoretical optimum because of internal irreversibilities resulting in entropy generation within the system and due to unwanted irreversible interactions with the surroundings of the system, the external irreversibilities (Franke, 1998).

To evaluate different energy technologies in such a manner that a comparison of different systems becomes possible, a method needs to be identified that can be systematically applied to all systems under consideration. A certain degree of generalization is consequently required to be able to compare technologies that fulfill a given task but do it in different ways. The basic generalization applied in energy engineering is the modeling of systems using energy and mass balances. The system under consideration is limited towards its surroundings by a theoretical boundary that can in principle be set arbitrarily but usually requires a good understanding of the considered system and of the goal of the analysis to be set in a sensible way. To allow a just comparison of different systems, a standardized approach to define boundaries should be chosen. This means that the definition of system boundaries should follow specified rules that are equal for all systems under consideration, otherwise not system performance but the choice of the system boundary could be the most significant influence on the results.

When the system boundary is defined in such a way, an analysis of the entering and exiting matter and energy flows can be performed. Since matter and energy are conserved the input flows are always balanced with the storage and output flows. The thermodynamic energy system evaluation can be based on a variety of thermodynamic properties (energy, entropy, exergy) and on a variety of different system boundary definitions. For example a system boundary can include all components that are required to produce and maintain the required system or be set in such a way that only its steady-state operation can be evaluated.

The most common thermodynamic assessment parameter is energy efficiency. It is considered an indicator how well an energy conversion or transfer process is accomplished (Cengel and Boles, 2006). However, this definition is not sufficiently precise, as energy efficiency does not take into account internal irreversibilities. Thus, a thermodynamic analysis based on energy only has the disadvantage that although it can identify energy losses, it is not sufficient to quantify the degree of irreversibility of a given process. Franke (1998) adds the consideration of entropy into the analysis to compensate this deficit. The problem of his method is that it allows only a close-to-process analysis and the optimization of a given process but is not well suited for cross-technology comparison. Consequently, neither the use of energy efficiency nor the entropy method are suitable as methods to compare different technology options universally and comprehensively.

A sophisticated, comprehensive and universal method suitable for thermodynamic energy system analysis is an assessment based on exergy. Exergy-based assessment takes into account the first and the second law of thermodynamics, it does not require reference processes and it is applicable to all types of processes (Fratzscher, 1997). The major disadvantage of an exergy-based evaluation is usually considered its dependence on reference state (Franke, 1998). However, the dependence of the exergetic performance evaluation of a thermodynamic system on the reference state reflects the dependence of the system operation on the conditions in its surroundings and should therefore be considered a sometimes inconvenient but necessary influence on thermodynamic technology evaluation.

2 The exergy-based transformability concept

This chapter covers the definition of the exergy-based properties transformability and transformation energy. Expressions for the transformability and transformation energy associated with most types of exergy are derived using reversible processes. The results are summarized in Tables A.2 on page 130 and A.3 on page 131. It becomes apparent that only those types of transformation energy that are based on a temperature difference to the environment are not fully transformable into other types of transformation energy. The chapter is concluded by the discussion of methods for calculating average transformabilities and effective thermal transformabilities that can be associated with differences of thermal transformation energy. The equations obtained for effective thermal transformabilities are summarized in Table A.4 on page 132.

The sign convention used for balances is system-centric. Flows entering a system are considered with a positive sign, while flows exiting are indicated by a negative sign. For defining equations, such as the definition equations for exergy flows or transformation energy flows, the algebraic sign indicates the direction of the defined flow in relation to the flow it is related to by the considered equation.

2.1 Introduction

It is commonly accepted that exergy is a useful property in engineering practice, which allows the determination of a thermodynamic value associated with a given flow. Exergy can be considered as a measure of quantity and quality, since its value is determined by the size of the considered flow as well as by its intensive properties. This characteristic distinguishes it clearly from energy, which has to be considered as a measure of quantity only. Furthermore, an exergy value can be associated with all types of flows while energy cannot be used to fully characterize *any* usable flow. Especially compressed gas flows at reference temperature and composition which can be associated with mechanical exergy have no significant difference of enthalpy in relation to a considered reference state. This discrepancy is caused by the low influence of pressure on enthalpy, which equals zero for ideal gases. Thus, if energy and exergy as means for flow assessment and characterization are compared, exergy appears to be the more universal and more comprehensive property. Nonetheless, exergy cannot substitute energy since it is a quasi property associated with a flow valid only for a given reference state. Additionally, the most significant advantage of exergy over energy, the fact that it also includes a quality aspect in addition to the quantity aspect of a given flow, has also a problematic aspect. The combination of quality and quantity aspects within one property makes it difficult to assess whether a given exergy value implies a high quality and a low quantity or a low quality and a high quantity. Lems et al. (2003) state for such a case where different aspects are merged into one that valuable information is lost, which in the case of exergy is the information to which extent quality effects have influenced it.

If a given exergy rate value is complemented with a matching value for mass or energy transfer, the quality of the flow should in principle become obvious. However, mass specific exergy can be defined only for mass transfers thus making it impossible to compare the quality of mass and mass-free energy flows directly. Additionally, a value of specific exergy cannot be interpreted as a measure for quality on its own since it only becomes meaningful in comparison to another specific exergy value. One of the major advantages of the exergy concept is its universality. Therefore, the question arises whether it is possible to separate the exergy associated with a transfer into a measure of quantity and a measure of quality, which are universal for energy and mass flows, intuitively understandable and independent of subjective choices of reference substances that are made in addition to the assumptions of the reference environment. It is hoped that with such a universal measure of quality the communication of the exergy concept and its implications to people not professionally occupied with thermodynamics can be significantly improved by providing an alternative perspective on exergy as a product of quality and quantity. Also, it could help to increase the precision of communication and deepen the understanding of exergy among professional engineers.

2.2 Splitting exergy into transformability and transformation energy for mass and energy flows

To be able to universally assess quality of a flow a property is sought that allows the assessment of quality associated with a flow on a dimensionless scale, that does not require reference values and that allows a quality assessment independently of the type of flow considered.

It has been stated by Szargut et al. (1988) that „the capacity of doing work has been accepted as a measure of the quality of energy". Therefore, an exergy-based measure of quality appears to be an ideal solution. It has already been indicated in subsection 1.2.1 on page 18 ff. that the use of a ratio of exergy rate to energy rate can be considered a first approach to define an exergy-based measure of quality. Since this approach is problematic if applied universally as has been discussed shortly in subsection 1.2.1 on page 18 ff., it seems that an universal exergy-based measure of flow quality has yet to be developed. Since the capacity of doing work has been accepted as a measure of quality, and could also be expressed as the transformability of a considered type of flow into work, it appears sensible to use the word *transformability* for the sought measure of quality. While the transformability should be the extract of the quality aspect, a matching property has to be defined that summarizes the quantity aspect of exergy in such a way that mass and energy flows remain universally comparable and evaluable.

To find a basis for the definition of such a property it appears reasonable to answer the question: What energetic aspect have all exergy flows in common? One plausible answer is: All exergy flows have in common that in order to generate work from the interaction of flow and environment, an energy flow into a conversion process is required that is equal or greater than the work generated. To complement transformability, this new exergy-based quantitative property will be labeled *transformation energy*.

2.2.1 Transformation energy

It has already been mentioned in subsection 1.1.6 on page 15 that using total exergy can be problematic in some cases. In order to allow a definition of exergetic efficiency for as many processes as possible, it appears sensible to define separate types of transformation energy that can in some cases be summed up to a total transformation energy, instead of defining such a total transformation energy from the start. The types of transformation energy should be defined in analogy to the matching types of exergy and will consequently be termed thermal, mechanical and chemical transformation energy.

In order to obtain the maximum amount of work, all exergy destruction and loss must be avoided. Only a reversible process allows the generation of work equal to the considered type of exergy, which is the potential to obtain work from an interaction of the considered flow and the environment. However, conservation of energy demands that in order to be able to generate the considered amount of work, the minimum energy input into such a reversible process must equal the work generated. If a flow contains less energy than is required for the generation of the maximum work, the lacking energy can only be obtained from the environment. Since the only energy contained in the environment is internal energy, the only transfer of energy available from the environment is heat at reference temperature. The energy input into a reversible process which is required to generate work equal to the relevant type of exergy (the potential to generate work) can be regarded as a measure of quantity for all types of exergy flows. This energy which is the sum of the energy input into the reversible process by the flow and the heat input from the environment, is minimally required to allow the transformation of the work potential associated with the flow (e.g. exergy) into actual work. Transformation energy can thus be defined as follows:

Transformation energy is the minimal amount of energy input into a set of reversible processes required to transform exergy into work.

Energy is transferred to the considered reversible process either by the considered flow or by heat transfer at reference temperature from the thermodynamic reference environment. For every type of exergy a corresponding type of transformation energy can be defined, which is a measure for the minimal energy required to transform the considered exergy type (mechanical, chemical, thermal...) into work.

In order to obtain expressions for the different types of transformation energy flows, appropriate reversible processes, which use the matching type of exergy flow, have to be identified. These reversible processes can then be used to derive type specific definitions for transformation energy. The derivation of transformation energy for the various types of exergy is discussed in sections 2.3 and 2.4.

The sign convention used for balances is system-centric. Flows entering the system are considered with a positive sign, while flows exiting are indicated by a negative sign.

Although transformation energy flows are defined using reversible processes they are like exergy flows associated to mass- and energy transfers in general. Since such mass- and energy transfers can enter or exit a thermodynamic system of interest, it does not appear sensible to use a general system-centric sign convention for defining equations. Therefore, for defining equations, such as the definition equations for exergy flows or transformation energy flows, the algebraic sign indicates the direction of the defined flow in relation to the flow it is related to by the considered equation. A positive sign indicates that the defined flow, e.g. exergy flow, and the flow it is related to, e.g. a heat flow, flow into the same direction. A negative sign consequently indicates that the defined flow is opposed to the flow it is related to, e.g. an exergy flow associated with a heat flow at a temperature below reference temperature is opposed to the exergy flow. In this case, if such a heat flow enters a system, it is associated with an exergy flow exiting the system, while if such a heat flow exits the system, it is associated with an exergy flow entering the system.

2.2.2 The compensation heat flow

In analogy to energy flows, mass flows and exergy flows balances can also be performed using transformation energy. Transformation energy as a type of energy should underlie the law of conservation, which means that it can neither be generated nor destroyed. Additionally, transformation energy should have the same algebraic sign or direction as the exergy flow that it is related to. This condition appears to be sensible since a ratio of exergy to transformation energy should provide a non-negative measure of quality.

The law of conservation is strictly valid only for mass and energy. As transformation energy does not always equal the energy of the flow it is associated with, since a part of the transformation energy can also originate in the environment, a special way to express transformation energy balances has to be developed.

It seems to be a sensible approach for the development of a set of rules for transformation energy balances to look for a way to compensate for the deviation of the transformation energy that is associated with a flow from the energy of that flow. A deviation of transformation energy from the energy of a flow implies that a part of the energy that is required to transform the potential into work (e.g. exergy) into actual work is obtained from the environment. Since the only type of energy that the environment can provide without limits is heat at reference temperature, this heat has to be considered in the transformation energy balance. The heat that needs to be associated with some types of energy and mass flows in addition to the transformation energy, can be termed *compensation heat* as it compensates the differences between transformation energy and the energy of the considered flow. This heat at reference temperature has not to be considered in the exergy balance since it is associated with an exergy value of zero. The compensation heat flow $\dot{Q}^*$ can be defined as the difference between the energy or enthalpy of the flow $\dot{E}n$ and the matching transformation energy flow $\dot{E}n_\tau$:

$$\dot{Q}^* = \dot{E}n - \dot{E}n_\tau \tag{2.1}$$

This equation is only valid if the reference state for the calculation of enthalpy is the same as the reference state used for exergy and transformation energy calculations.

Using this definition, the transformation energy balance of a simple open system can be expressed as a function of transformation energy flows and compensation heat flows entering (subscript i) or exiting (subscript e) the system boundary:

$$0 = \dot{E}n_{\tau,i} + \dot{Q}^*_i - \dot{E}n_{\tau,e} - \dot{Q}^*_e$$

2 The exergy-based transformability concept

Which equals the energy balance:
$$0 = \dot{E}n_i - \dot{E}n_e$$

Since for different types of mass or energy flows different definitions of transformation energy apply, different compensation heat flows have to be considered for each type of transformation energy in a transformation energy balance. Thus, in addition to deriving expressions for the different types of transformation energy, expressions for the compensation heat flow have to be derived to provide the basis for the application of the transformation energy balance. A discussion of the transformation energy balance can be found in section 3.1 on page 48 ff.

2.2.3 Transformability

By defining transformation energy, a universal measure of quantity has been found. It is now simple to define transformability τ in analogy to the exergy to energy ratio as the ratio of the exergy rate $\dot{E}$ to the transformation energy rate $\dot{E}n_\tau$ associated with a flow:

$$\tau = \frac{\dot{E}}{\dot{E}n_\tau} \tag{2.2}$$

Transformability can be viewed as a measure of the thermodynamic quality of the mass or energy flow under consideration. It relates exergy as a measure for quantity and quality to transformation energy as an equally universal measure of quantity - leaving it to be a dimensionless measure of quality only. The common basis for the calculation of exergy and transformation energy is the combined system of considered flow and reference environment. Transformability is therefore more universal than the exergy rate to energy rate ratio since it relates exergy not only to a property of the considered flow but to a property of the combined system. Transformability can only have values between $0\,\%$ and $100\,\%$. Since transformability is a function of two potentials (exergy and transformation energy) the transformability is a potential also. Transformability is an intensive property that can be associated with any transformation energy flow for any given reference state.

A drawback of a quality measure that relates exergy to the required energy input into a reversible system is its comparably high level of complexity. It appears probable that the novel concept can only be understood fully after an in depth review. In spite of that, if transformability is accepted as a measure of thermodynamic quality without deeper understanding, the characterization of the quality of energy and mass transfers becomes straightforward, thermodynamically correct and easily communicable. To allow a direct use of transformability and transformation energy, the expressions of the properties and the matching compensation heat flows that are derived in the following section are tabulated in Table A.2 on page 130 and Table A.3 on page 131.

2.3 Transformation energy and transformability associated with conductive heat flows

Conductive heat flows are transfers of kinetic energy from molecules to adjacent molecules without movement of the substance as a whole. The reversible process that allows the conversion of heat into work is an ideal heat engine that operates between a reservoir at the flow temperature T and a reservoir at reference temperature T_0. Since the transformation energy is the input into the work generating process, it is always entering the ideal heat engine at the higher temperature - at T for heat flows above reference temperature and at T_0 if heat flows below reference temperature are evaluated. Since the input into the reversible heat engine differs depending on the relation of flow temperature to reference temperature, it is necessary to consider heat flows with temperatures above and heat flows with temperatures below reference temperature separately.

The flows are labelled according to Figure 2.1. The dashed line symbolizes the balance boundary.

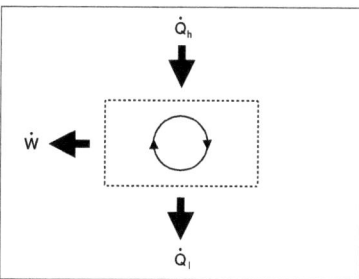

Figure 2.1: Energy flow chart of a reversible heat engine process

In the following, an expression for conductive heat flows at a temperature T, which is larger than reference temperature T_0, is derived. The energy balance of a reversible heat engine can be expressed as a function of a heat flow $\dot{Q}_h$ at a high temperature T_h, a heat flow $\dot{Q}_l$ at a low temperature T_l and the generated work flow $\dot{W}$:

$$0 = \dot{Q}_h - \dot{Q}_l - \dot{W} \qquad (2.3)$$

For $T \geq T_0$: $T_l = T_0$

In order to generate work from the considered heat flow at temperature $T = T_h$, the total minimally required energy input equals the heat input into the heat engine. Thus, the transformation energy associated with conductive heat flows $\dot{En}_\tau^Q$ above reference temperature equals the considered conductive heat flow:

$$\dot{En}_\tau^Q = \dot{Q}_h \qquad (2.4)$$

For $T \geq T_0$ and a heat flow originating from a reservoir at a constant temperature T, the associated exergy rate $\dot{E}^Q$ is defined as (Moran and Shapiro, 2007):

$$\dot{E}^Q = \left(1 - \frac{T_0}{T}\right) \cdot \dot{Q} \qquad (2.5)$$

According to the definition of transformability in Equation 2.2 and considering that $\dot{Q} = \dot{Q}_h$, the transformability associated with a conductive heat flow τ^Q can be expressed as :

$$\tau^Q = \frac{\dot{E}^Q}{\dot{En}_\tau^Q} = 1 - \frac{T_0}{T} \qquad (2.6)$$

A compensation heat flow is not associated with conductive heat flows at $T \geq T_0$ since by using Equation 2.1 the following expression is obtained:

$$\dot{Q}^* = \dot{Q}_h - \dot{En}_\tau = 0$$

For conductive heat flows above reference temperature the transformability consequently equals the exergy rate to energy rate ratio, which has been proposed as a measure of quality in literature. The transformation energy associated with a conductive heat flow above reference temperature is identical with the heat flow and no compensation energy has to be introduced into the transformation energy balance.

For $T < T_0$, the considered heat flow is $\dot{Q}_l$, while the energy input still equals $\dot{Q}_h$. As transformation energy is associated with $\dot{Q}_l$, the heat input has to be expressed as a function of $\dot{Q}_l$. For a reversible cycle the following relation is valid (Moran and Shapiro, 2007):

$$\frac{\dot{Q}_h}{\dot{Q}_l} = \frac{T_h}{T_l}$$

Thus, the absolute value of the input heat at reference temperature can be expressed as:

$$\dot{Q}_0 = \dot{Q}_h = \frac{T_0}{T} \cdot \dot{Q}_l \qquad (2.7)$$

The transformation energy associated with the considered flow must consequently have the same absolute value as the heat from the environment. However, one aspect of transformation energy is different from the heat flow at reference temperature. In the transformation energy balance, the transformation energy has to substitute the heat flow it is associated with, which in this case is $\dot{Q}_l$. While the transformation energy flow enters the considered process like $\dot{Q}_h$, the heat flow it is associated with exits the process. Thus, to properly define transformation energy in accord with its definition, the heat at reference temperature in Equation 2.7 has to be prefaced by a negative sign to provide the definition of transformation energy. As a consequence for temperatures below reference temperature, the transformation energy associated with conductive heat flows is calculated in relation to the low temperature heat flow $\dot{Q} = \dot{Q}_l$ as:

$$\dot{En}_\tau^Q = -\frac{T_0}{T} \cdot \dot{Q}_l \qquad (2.8)$$

With this expression and the exergy definition from Equation 2.5 in which in this case $\dot{Q} = \dot{Q}_l$, the transformability associated with a conductive heat transfer at a temperature below reference

2 The exergy-based transformability concept

temperature is calculated as:

$$\begin{aligned} \tau^Q &= \frac{\left(1 - \frac{T_0}{T}\right) \cdot \dot{Q}_l}{-\left(\frac{T_0}{T}\right) \cdot \dot{Q}_l} \\ &= -\frac{T - T_0}{T} \cdot \frac{T}{T_0} \\ &= -\frac{T - T_0}{T_0} \\ &= 1 - \frac{T}{T_0} \end{aligned} \qquad (2.9)$$

The transformability associated with flows at temperatures below reference temperature is in principle similar to the transformability associated with a high temperature heat flow expressed in Equation 2.6 since in both cases it is calculated as:

$$\tau^Q = 1 - \frac{T_l}{T_h}$$

This expression equals the efficiency of a reversible heat engine or the so called Carnot factor (Moran and Shapiro, 2007). As a measure of quality, it provides a direct expression of the share of the considered heat input into a reversible process that can be converted to power. The compensation heat flow associated with a conductive heat flow at a temperature below reference temperature can be obtained using Equations 2.1 and 2.8 as:

$$\begin{aligned} \dot{Q}_l^* &= \dot{Q}_l - \dot{E}n_\tau \\ &= \dot{Q}_l - \left(-\dot{Q}_l \cdot \frac{T_0}{T}\right) \\ &= \left(1 + \frac{T_0}{T}\right) \cdot \dot{Q}_l \end{aligned} \qquad (2.10)$$

This compensation heat flow is larger than the heat flow it is associated with. Since for the evaluation of energy systems and system analysis only the difference of compensation heat flows or the so called effective compensation heat flow is relevant, the absolute value of this flow has no deeper meaning. See subsection 3.2.3 on page 54 ff. for a discussion.

Figure 2.2 shows the exergy flow chart and the transformation energy flow chart of a reversible power cycle operating between thermal reservoirs at $T < T_0$ and T_0.

It becomes obvious that the transformation energy flow diagram increases the complexity of the balance for the considered system. The higher complexity of the diagram in comparison to the exergy flow chart or the energy flow chart in Figure 2.1 can be considered the price for the higher transparency that is possible by dividing exergy into transformation energy and transformability. For a more detailed discussion of the transformation energy balance see subsection 3.1 on page 48 ff.

The obtained results are valid only for conductive heat flows that are provided at constant temperature,

2 The exergy-based transformability concept

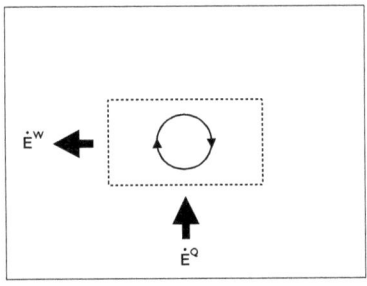
(a) Exergy flow chart

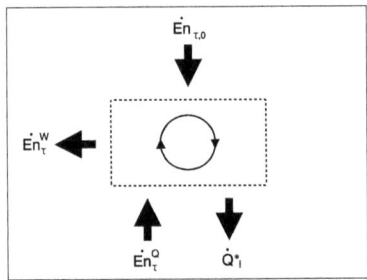
(b) Transformation energy flow chart

Figure 2.2: Exergy and transformation energy flow charts of a reversible power cycle operating between reference temperature and a temperature below reference temperature

which could originate for example by an ideal heat exchanger surface connected to a combustion chamber or be part of an idealized condenser.

2.4 Transformation energy and transformability associated with mass flows

In contrast to mass-free energy transfers which are always associated with only one type of exergy mass flows can be associated with a variety of exergy types such as mechanical, thermal or chemical exergy. In analogy to exergy evaluation and analysis which becomes more accurate if the types of exergy are considered separately (Lazzareto and Tsatsaronis, 2006, p. 1260), the matching types of transformation energy should also be calculated separately in order to obtain the most accurate evaluation of the considered flows.

2.4.1 Thermal transformation energy and transformability associated with mass flows

Thermal exergy associated with a mass flow is evaluated along an isobaric line and is a measure for the maximum work obtainable by bringing the mass flow from its initial state (T, p) to a thermal equilibrium with the environment at (T_0, p) (Lazzareto and Tsatsaronis, 2006). The reversible process that allows the generation of work from the difference in temperature between flow and environment is the same as the one that allows the generation of work from conductive heat flows: the reversible heat engine. The difference to the evaluation of conductive heat flows is that the exergy transfer is associated with mass flows thus, instead of a conductive heat flow at a constant temperature, the relevant heat transfer is the one resulting from a specific enthalpy difference between flow input and exit $(h_i - h_e)$ of the considered mass flow $\dot{m}$. In order to avoid confusion between these two types of energy transfer, the superscript H is used to signify that the considered heat flow $\dot{Q}^H$ is the result of an enthalpy difference between a mass flow at the inlet and at the exit. This heat flow $\dot{Q}^H$ can thus be expressed as:

$$\dot{Q}^H = \dot{m} \cdot (h_i - h_e) \qquad (2.11)$$

2 The exergy-based transformability concept

Adapting the nomenclature accordingly, the energy balance of an ideal heat engine as shown in Figure 2.1 can be expressed for heat flows $\dot{Q}_h^H$ originating from mass flows at temperatures above reference temperature as:

$$0 = \dot{Q}_h^H - \dot{Q}_l - \dot{W}$$

The total energy input into a reversible heat engine is $\dot{Q}_h^H$. If this heat flow is a result of the cooling of a mass flow from initial to reference temperature at constant pressure, this heat flow and consequently the thermal transformation energy can be expressed as:

$$\dot{En}_\tau^T = \dot{m} \cdot (h - h_{T0}) = \dot{Q}_h^H \qquad (2.12)$$

In this equation $h = h(T, p)$ and $h_{T0} = h(T_0, p)$ where p the pressure of the flow is kept constant during the heat exchange. The chemical composition also remains unchanged.

In Equation 1.1 thermal exergy has been defined as a function of specific enthalpy, reference temperature and specific entropy s of the flow, which, using the introduced abbreviations, can be expressed as:

$$\dot{E}^T = \dot{m} \cdot [h - h_{T0} - T_0 \cdot (s - s_{T0})] \qquad (2.13)$$

Thus, for mass flows associated with thermal exergy at $T \geq T_0$, the thermal transformability τ^T can be defined based on Equations 2.2, 2.4 and 2.13 as:

$$\tau^T = 1 - \frac{T_0 \cdot (s - s_{T0})}{(h - h_{T0})}$$

In thermodynamics, for temperature changes of mass flows at constant pressure, it is common to define a thermodynamic average temperature T_a as (Bejan et al., 1996) [1]:

$$T_a = \frac{h_e - h_i}{s_e - s_i} \qquad (2.14)$$

Defining the thermodynamic average temperature between considered flow temperature and reference temperature as:

$$T_{a0} = \frac{h - h_{T0}}{s - s_{T0}}$$

, thermal transformability associated with mass flows at $T \geq T_0$ can be defined as:

$$\tau^T = 1 - \frac{T_0}{T_{a0}}$$

---

[1] For ideal gas flows with a constant specific heat capacity at a constant pressure, the thermodynamic average temperature equals the logarithmic mean temperature (Fratzscher et al., 1986):

$$T_a = \frac{T_i - T_e}{\ln \frac{T_i}{T_e}}$$

2 The exergy-based transformability concept

The compensation heat flow for this type of transformation energy flows is calculated using the expressions from Equations 2.11, 2.12 and 2.1 as:

$$\begin{aligned} \dot{Q}^* &= \dot{Q}^H - \dot{E}n_\tau^T \\ &= \dot{m} \cdot [h - h_{T0} - (h - h_{T0})] \\ &= 0 \end{aligned}$$

Like transformability associated with conductive heat flows above reference temperature, thermal transformability of mass flows with a temperature above reference temperature equals the energy rate to exergy rate ratio.

For mass flows at temperatures below reference temperature, thermal exergy is defined in Equation 2.13. Figure 2.3 shows the mass and the energy flows in a reversible heat engine used for the assessment of thermal transformation energy associated with a mass flow below reference temperature. The reference state for the enthalpy calculation is equal to the reference state of exergy so that the considered heat flow $\dot{Q}_l^H$ equals the enthalpy flow $\dot{H}$:

$$\dot{Q}_l^H = \dot{H}$$

Since the specific enthalpy of the mass flow is lower than the specific enthalpy of the mass at reference conditions, the enthalpy flow has a negative sign which indicates a direction opposed to the mass flow.

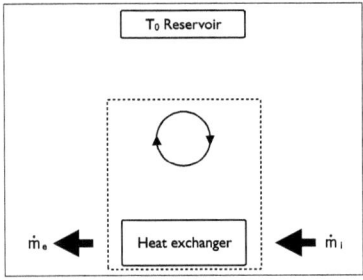
(a) Mass flow chart

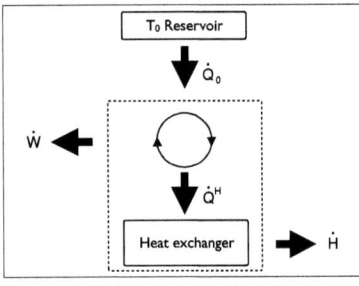
(b) Energy flow chart

Figure 2.3: Energy and mass flow charts of a reversible power cycle used for transformation energy derivation for mass flows at temperatures below reference temperature

The energy balance according to Figure 2.3 can be expressed as:

$$0 = \dot{Q}_0 - \dot{Q}^H - \dot{W}$$

The definition of the thermodynamic average temperature for constant pressures (Equation 2.14) can be transposed to give:

$$T_{a0} \cdot (s - s_{T0}) = (h - h_{T0})$$

2 The exergy-based transformability concept

Assuming a constant mass flow through a heat exchanger, the low temperature heat flow can be expressed as:

$$\dot{Q}^H = \dot{m} \cdot (h - h_{T0}) = \dot{m} \cdot T_{a0} \cdot (s - s_{T0}) \qquad (2.15)$$

For a reversible cycle the following expression is valid (Moran and Shapiro, 2007):

$$\frac{\dot{Q}_h}{\dot{Q}_l} = \frac{T_h}{T_l}$$

If a reversible cycle obtains a conductive heat flow from the environment and discharges heat to a mass flow at constant pressure the following relation is valid:

$$\frac{\dot{Q}_0}{\dot{Q}^H} = \frac{T_0}{T_{a0}} \qquad (2.16)$$

Using Equations 2.16 and 2.11, the heat at reference temperature which equals the absolute value of the thermal transformation energy can be expressed as a function of the considered mass flow $\dot{m}$.

$$\begin{aligned}
\dot{Q}_0 &= \frac{T_0}{T_{a0}} \cdot \dot{Q}^H \\
&= \frac{T_0}{T_{a0}} \cdot \dot{m} \cdot T_{a0} \cdot (s - s_{T0}) \\
&= \dot{m} \cdot T_0 \cdot (s - s_{T0}) \\
&= -\dot{E}n_\tau^T \qquad (2.17)
\end{aligned}$$

In analogy to the evaluation of conductive heat flows, the association of transformation energy with a mass flow at a temperature below reference temperature requires the introduction of a negative sign into the relation of heat flow and transformation energy rate.

For thermal transformability τ^T associated with the considered mass flow, using Equations 2.12 and 2.13, the following expression is obtained:

$$\begin{aligned}
\tau^T &= \frac{\dot{E}^T}{\dot{E}n_\tau^T} \\
&= \frac{e^T}{en_\tau^T} \\
&= \frac{h - h_{T0} - T_0 \cdot (s - s_{T0})}{-T_0 \cdot (s - s_{T0})} \\
&= \frac{h - h_{T0}}{-T_0 \cdot (s - s_{T0})} + 1 \\
&= \frac{T_{a0} \cdot (s - s_{T0})}{-T_0 \cdot (s - s_{T0})} + 1 \\
&= 1 - \frac{T_{a0}}{T_0}
\end{aligned}$$

Thus, the thermal transformability associated with low temperature mass flows equals the transformability associated with low temperature conductive heat flows at $T = T_{a0}$.

2 The exergy-based transformability concept

The thermal compensation heat flow for thermal transformation energy can be calculated based on Equations 2.1, 2.11, 2.12 and 2.14 as:

$$\begin{aligned} \dot{Q}^{*T} &= \dot{Q}^H - \dot{E}n_\tau^T \\ &= \dot{m} \cdot [h - h_{T0} + T_0 \cdot (s - s_{T0})] \\ &= \dot{Q}^H \cdot (1 + \frac{T_0}{T_{a0}}) \end{aligned}$$

It becomes apparent that the compensation heat flow associated with a mass flow at a temperature below reference temperature is defined analogously to the compensation heat flow associated with a conductive heat flow at a temperature below average temperature which has been presented in Equation 2.10.

2.4.2 Mechanical transformation energy and transformability associated with mass flows

A reversible process suitable for determination of mechanical transformation energy is the reversible expansion process along an isothermal line at reference temperature T_0. If the pressure of the mass flow is greater than the reference pressure, the input into this process is a mass flow at reference temperature $\dot{m}_i(T_0, p)$ which is being discharged as $\dot{m}_e(T_0, p_0)$, while retaining its chemical composition.

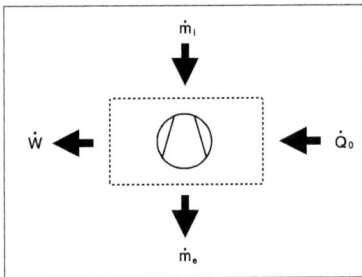

Figure 2.4: Energy and mass flow chart of a reversible heat engine process

The energy balance of the reversible expansion process for a mass flow can be expressed using the labels in Figure 2.4 as:

$$\begin{aligned} 0 &= \dot{m}_i \cdot h_i - \dot{m}_e \cdot h_e - \dot{W} + \dot{Q}_0 \\ &= \dot{m}_i \cdot (h_i - h_e) - \dot{W} + \dot{Q}_0 \end{aligned} \quad (2.18)$$

It is important to keep in mind that mechanical transformation energy can only be evaluated at T_0 since only at that temperature thermal energy from the environment can be provided without limit.

2 The exergy-based transformability concept

Heat from the environment is required to compensate for the potential temperature decrease of the considered mass flow to allow an expansion along an isothermal line.

The difference $(h_i - h_e)$ can be either positive or negative depending on the chemical composition of the flow evaluated. This makes it rather difficult to determine whether the heat flow associated with $\dot{m}_i$:

$$\dot{Q}_i^H = \dot{m}_i \cdot (h_i - h_e) \qquad (2.19)$$

is an input or an output flow of the considered expansion process. For this reason Figure 2.4 shows mass-free energy transfers and mass flows instead of enthalpy flows.

To find expressions for the total energy input into the reversible process, both cases can be considered separately. If $(h_i - h_e) \geq 0$, then the enthalpy flow associated with $\dot{m}_i$ is an input. The total energy input and thus the mechanical transformation energy $\dot{E}n_\tau^M$ associated with the considered mass flow can be expressed as:

$$\dot{E}n_\tau^M = \dot{m}_i \cdot (h_i - h_e) + \dot{Q}_0 \qquad (2.20)$$

If $(h_i - h_e) < 0$, the enthalpy flow is an energy output. However, since this energy is transferred at reference temperature, it can be added to the conductive heat input at reference temperature $\dot{Q}_0$ to provide an equation for the net heat input at reference temperature[2]. Thus, the mechanical transformation energy can again be expressed as:

$$\dot{E}n_\tau^M = \dot{m}_i \cdot (h_i - h_e) + \dot{Q}_0$$

The direction of the enthalpy flow associated with $\dot{m}_i$ has therefore no influence on the definition of mechanical transformation energy.

If a mass flow above reference temperature is isothermally expanded, its specific entropy changes from the state $s_{T0}(T_0, p)$ to state $s_0(T_0, p_0)$. Entropy change for a reversible thermodynamic cycle is given by Moran and Shapiro (2007) as:

$$dS = \left(\frac{dQ}{T}\right)_{rv}$$

This equation can be transposed to give:

$$Q_{rv} = \int T dS$$

If a heat flow is transferred at a constant temperature of the environment to a steady-state mass flow, this heat flow can be expressed as:

$$\dot{Q}_0 = \dot{m}_i \cdot T_0 \cdot (s_e - s_i) \qquad (2.21)$$

---

[2] Since transformation energy is defined as the *minimal* energy input into a reversible process to generate work equal to the considered exergy, all heat inputs at reference temperature are considered as a sum. This net heat input into the process is the minimal heat flow from the reference environment necessary to transform exergy into actual work.

2 The exergy-based transformability concept

Assuming that pressure p of the mass flow is larger than reference pressure p_0, the specific entropies can be expressed as $s_{T0} = s_i < s_e = s_0$, which leads to:

$$\dot{Q}_0 = -\dot{m}_i \cdot T_0 \cdot (s_{T0} - s_0) \qquad (2.22)$$

Using this expression for the conductive heat flow from the environment, the mechanical transformation energy flow given by Equation 2.20 can be expressed using $h_i = h_{T0} = h(T_0, p)$, $h_e = h_0 = h(T_0, p_0)$ and $\dot{m} = \dot{m}_e$ as:

$$\dot{En}_\tau^M = \dot{m} \cdot [h_{T0} - h_0 - T_0 \cdot (s_{T0} - s_0)] \qquad (2.23)$$

Since a mechanical exergy flow $\dot{E}^M$ associated with $\dot{m}$ can be obtained from Equation 1.2 as:

$$\dot{E}^M = \dot{m} \cdot [h_{T0} - h_0 - T_0 \cdot (s_{T0} - s_0)] \qquad (2.24)$$

the mechanical transformability τ^M associated with compressed mass flows at $p \geq p_0$ is defined as:

$$\tau^M = \frac{\dot{E}^M}{\dot{En}_\tau^M} = 1$$

To be able to integrate mechanical transformation energy flows into a transformation energy balance, the compensation heat flow needs to be determined. For mechanical transformation energy associated with flows above reference pressure, it can be calculated based on Equations 2.1, 2.19 and 2.20 as:

$$\begin{aligned}\dot{Q}^{*,M} &= \dot{Q}_i^H - \dot{En}_\tau^M \\ &= \dot{m}_i \cdot \{(h_i - h_e) - [(h_i - h_e) - T_0 \cdot (s_i - s_e)]\} \\ &= \dot{m} \cdot T_0 \cdot (s_{T0} - s_0)\end{aligned} \qquad (2.25)$$

For mass flows at $p < p_0$ the mass flow with which exergy is associated is $\dot{m}_e$ from Figure 2.4. Since it is always the mass flow with the higher pressure that enters the expansion process, in this case $\dot{m}_i$ equals a mass inflow at reference pressure and conditions which is expanded to the considered conditions of the exiting mass flow $\dot{m}_e$. For gases enthalpy is only a weak function of pressure, so that for low pressures $h \approx h(T)$ and $s > s_0$. Consequently, the specific mechanical exergy (see Equation 1.2) associated with a mass flow below reference pressure is negative. Bosnjakovic and Knoche (1998) have noted that processes generating mass flows from the environment can also result in a generation of work. As a consequence, a negative sign associated with the specific exergy can consistently be interpreted as a sign that the mechanical exergy flow is opposed to the direction of the mass flow it is associated with. A consistent interpretation of the negative sign of specific exergy and transformation energy is necessary to balance these flows correctly.

The energy balance for a reversible expansion process expanding matter isothermally at T_0 from reference pressure to the target conditions below reference pressure remains the same as given in Equation 2.18. The difference is that now matter from the environment is expanded instead of

2 The exergy-based transformability concept

matter of the considered flow and that exergy is associated with $\dot{m}_e$ instead of $\dot{m}_i$. Since a negative sign indicates a direction in relation to the considered flow $\dot{m}_i = -\dot{m}_e$, the energy balance can be expressed as:

$$\begin{aligned} 0 &= \dot{m}_i \cdot h_i - \dot{m}_e \cdot h_e - \dot{W} + \dot{Q}_0 \\ &= -\dot{m}_e \cdot (h_i - h_e) - \dot{W} + \dot{Q}_0 \end{aligned}$$

Since Equation 2.21 gives a definition for the heat flow at reference pressure in relation to $\dot{m}_i$, but the flow to which exergy is associated is $\dot{m}_e$, the equation needs to be adapted accordingly.

$$\begin{aligned} \dot{Q}_0 &= \dot{m}_i \cdot T_0 \cdot (s_e - s_i) \\ &= -\dot{m}_e \cdot T_0 \cdot (s_e - s_i) \end{aligned}$$

The total energy input into a reversible expander and consequently the mechanical transformation energy associated with a mass flow below reference pressure can therefore be calculated based on the general definition in Equation 2.20 as:

$$\begin{aligned} \dot{En}_\tau^M &= \dot{m}_i \cdot (h_i - h_e) + \dot{Q}_0 \\ &= -\dot{m}_e \cdot (h_i - h_e) - T_0 \cdot (s_e - s_i) \\ &= \dot{m}_e \cdot [h_e - h_i - T_0 \cdot (s_e - s_i)] \end{aligned}$$

Introducing $\dot{m} = \dot{m}_e$, index i = index 0 (indicating reference temperature and pressure) and index e = index $T0$ (indicating nonreference pressure at reference temperature) the mechanical transformation energy associated with mass flows at pressures below reference pressure can be calculated as:

$$\dot{En}_\tau^M = \dot{m} \cdot [h_{T0} - h_0 - T_0 \cdot (s_{T0} - s_0)]$$

Since this equation is the same as Equation 2.23 it has been shown that it is valid for mechanical transformation energy in general.

As Equation 2.24 is also generally valid for mechanical exergy, the transformability associated with flows below reference pressure equals the one for flows above reference pressure:

$$\tau^M = \frac{\dot{E}^M}{\dot{En}_\tau^M} = 1$$

Finally, the mechanical compensation heat flow that is required to fulfill the energy balance is

2 The exergy-based transformability concept

calculated based on Equations 2.1, 2.19 and 2.20 as:

$$\begin{aligned} \dot{Q}^{*,M} &= \dot{Q}_i^H - \dot{E}n_\tau^M \\ &= -\dot{m}_e \cdot \{(h_i - h_e) - [(h_i - h_e) - T_0 \cdot (s_i - s_e)]\} \\ &= \dot{m} \cdot T_0 \cdot (s_{T0} - s_0) \end{aligned}$$

Since the final expression equals the one from Equation 2.25, it has been shown that this expression for the mechanical compensation heat flow is valid for all pressures.

Concluding, it can be summarized that the mechanical transformation energy and mechanical transformability for mass flows at all pressures is calculated by one set of equations. The principal difference of mass flows at $p < p_0$ from mass flows at $p \geq p_0$ is, therefore, the negative specific mechanical exergy and transformation energy that is associated with these flows. According to the used sign convention this can be consistently interpreted as an indicator that the considered exergy or transformation energy flow has the opposite direction in relation to the mass flow it is associated with.

2.4.3 Reactive transformation energy and transformability associated with mass flows

Reactive transformation energy is the energy input into a reversible fuel cell which allows the generation of work based on a reaction of the considered fuel with molecules found in the environment. The specific energy and exergy flow schemes for the determination of reactive transformation energy are shown in Figure 2.5.

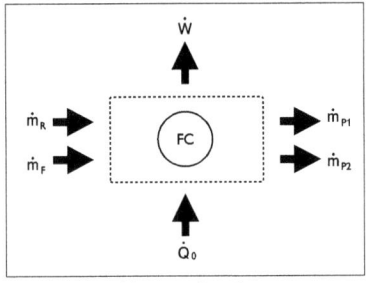
(a) Energy flow chart

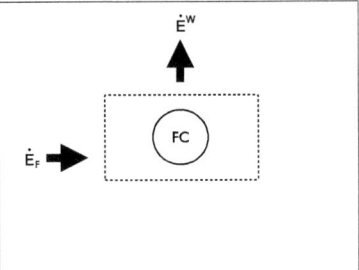
(b) Exergy flow chart

Figure 2.5: Flow charts of a reversible fuel cell process used for reactive transformability derivation

The definition of reactive transformation energy and transformability requires knowledge of a standard reaction of the fuel with components of the environment to products present in the environment. Therefore, it is convenient to express the equations on a molar basis. The labels fuel and product are used in this context for the fuel and the products of the considered reversible fuel cell.

The energy balance of a reversible isothermal fuel cell where the substance of interest and all reactands and products enter or exit at reference conditions (T_0, p_0) can be expressed as a function of the molar

2 The exergy-based transformability concept

flow rate of the fuel $\dot{n}_F$, the molar specific enthalpies of formation $\bar{h}_f$ of fuel (subscript F), reactand (subscript R) and product (subscript P), their stoichiometric coefficients ν and the heat input at reference temperature $\dot{Q}_0$ (Moran and Shapiro, 2007) :

$$0 = \dot{n}_F \cdot \left[\bar{h}_{fF} + \sum_R \frac{\nu_R}{\nu_F} \cdot \bar{h}_{fR} - \sum_P \frac{\nu_P}{\nu_F} \cdot \bar{h}_{fP} \right] - \dot{W} + \dot{Q}_0$$

Since all inputs and outflows are at reference temperature and pressure, only the net input of heat from the environment has to be considered. Thus, the reactive transformation energy $\dot{En}_{\tau,F}^R$ equals the sum of the heat and enthalpy input into the fuel cell:

$$\dot{En}_{\tau,F}^R = \dot{n}_F \cdot \left[\bar{h}_{fF} + \sum_R \frac{\nu_R}{\nu_F} \cdot \bar{h}_{fR} - \sum_P \frac{\nu_P}{\nu_F} \cdot \bar{h}_{fP} \right] + \dot{Q}_0 \qquad (2.26)$$

An equation for $\dot{Q}_0$ as a function of specific entropy can be obtained based on the following entropy equation, which is valid for reversible thermodynamic cycles Moran and Shapiro (2000):

$$\left(\frac{\dot{Q}}{\dot{m}} \right)_{rv} = \int_1^2 T ds$$

A heat transfer to or from a mass flow that does not change the temperature of the mass flow must change its entropy. Naturally, reactions occur only accompanied with entropy increase so that the sum of the specific absolute entropies[3] of fuels and reactands is smaller than the sum of absolute entropies of the products. Considering Figure 2.5 the heat flow enters the process and is therefore positive. A negative sign has to be added here to correctly relate the entropy change in the reaction, resulting in a net entropy output, to the heat input into the reversible fuel cell. For a reaction to which heat at reference temperature $\dot{Q}_0$ is transferred, this equation could be specified as a function of absolute molar specific entropies $\bar{s}^o$ as:

$$\dot{Q}_0 = -\dot{n}_F \cdot \int_P^{F,R} T_0 d\bar{s}^o$$

Solving the integral, heat from the environment entering a reversible fuel cell can be expressed as :

$$\dot{Q}_0 = -\dot{n}_F \cdot \left[T_0 \bar{s}^o_F + \sum_R \frac{\nu_R}{\nu_F} \cdot T_0 \bar{s}^o_R - \sum_P \frac{\nu_P}{\nu_F} \cdot T_0 \bar{s}^o_P \right] \qquad (2.27)$$

The reactive transformation energy can thus be specified based on Equations 2.26 and 2.27 as:

$$\dot{En}_{\tau,F}^R = \dot{n}_F \cdot \left[(\bar{h}_{fF} - T_0 \bar{s}^o_F) + \sum_R \frac{\nu_R}{\nu_F} \cdot (\bar{h}_{fR} - T_0 \bar{s}^o_R) - \sum_P \frac{\nu_P}{\nu_F} \cdot (\bar{h}_{fP} - T_0 \bar{s}^o_P) \right] \qquad (2.28)$$

---
[3] Chemical exergy is defined using absolute entropy values, i.e. entropy values that represent the entropy change of a substance taken from absolute zero to a given temperature.

2 The exergy-based transformability concept

A reactive exergy flow $\dot{E}_F^R$ associated with a fuel flow can be calculated by (see appendix A2 on page 124 ff.):

$$\dot{E}_F^R = \dot{n}_F \cdot \left[(\bar{h}_{fF} - T_0\bar{s}^o{}_F) + \sum_R \frac{\nu_R}{\nu_F} \cdot (\bar{h}_{fR} - T_0\bar{s}^o{}_R) - \sum_P \frac{\nu_P}{\nu_F} \cdot (\bar{h}_{fP} - T_0\bar{s}^o{}_P) \right]$$

Thus, for fuel flows reacting with the environment, the reactive transformability τ^R is calculated as:

$$\tau_F^R = \frac{\dot{E}_F^R}{\dot{E}n_{\tau,F}^R} = 1$$

The reactive compensation heat flow, associated with the considered mass flow is calculated, based on Equations 2.1 and 2.28, as a function of the higher heating value flow $H\dot{H}V$, the net energy transfer to the process by substance flows:

$$\begin{aligned}
\dot{Q}^{*,R} &= H\dot{H}V - \dot{E}n_\tau^R \\
&\quad - \dot{n}_F \cdot \left[\bar{h}_{fF} + \sum_R \frac{\nu_R}{\nu_F} \cdot \bar{h}_{fR} - \sum_P \frac{\nu_P}{\nu_F} \cdot \bar{h}_{fP} \right] \\
&\quad - \dot{n}_F \cdot \left[(\bar{h}_{fF} - T_0\bar{s}^o{}_F) + \sum_R \frac{\nu_R}{\nu_F} \cdot (\bar{h}_{fR} - T_0\bar{s}^o{}_R) - \sum_P \frac{\nu_P}{\nu_F} \cdot (\bar{h}_{fP} - T_0\bar{s}^o{}_P) \right] \\
&= \dot{n}_F \cdot \left[T_0\bar{s}^o{}_F + \sum_R \frac{\nu_R}{\nu_F} \cdot T_0\bar{s}^o{}_R - \sum_P \frac{\nu_P}{\nu_F} \cdot T_0\bar{s}^o{}_P \right]
\end{aligned}$$

The absolute value of this expression equals the absolute value of the heat flow at reference conditions given in Equation 2.27 that is required to perform the reversible reaction. According to Bejan et al. (1996), in technical literature chemical exergy of fuels is often approximated with the higher heating value of the considered fuel. I.e. the difference between higher heating value (HHV) and chemical exergy of dry ashfree coal is given in the same source as being approximately 2 %. For practical and exemplary evaluations it appears to be sufficient to consider the higher heating value of fuels and chemical exergy as equal. As transformation energy equals chemical exergy for general calculations, this approach can be extended to the transformability evaluation. As a consequence of such a simplification, the reactive chemical compensation heat flow would become negligible.

2.4.4 Nonreactive transformation energy and transformability associated with mass flows

Nonreactive exergy E^N is associated with concentration differences between flow and environment of substances present in the environment if evaluated at reference temperature and pressure. Tsatsaronis (2007) defines it as being associated with nonreactive processes such as expansion, compression, mixing and separation. It also has to be considered for the reactands and products when determining the chemical exergy associated with a substance not present in the reference environment. The chemical exergy flow $\dot{E}^{CH}$ associated with a fuel flow can in principle be expressed as a function of

2 The exergy-based transformability concept

the reactive exergy associated with the fuel as well as the nonreactive exergies associated with pure product and reactand flows, which are assumed in the calculation of reactive exergy of fuels (Moran and Shapiro, 2007):

$$\dot{E}_F^{CH} = \dot{E}_F^R + \sum \dot{E}_R^N - \sum \dot{E}_P^N$$

The reversible process suitable for the determination of nonreactive exergy is the reversible fuel cell that can also be used for the determination of reactive transformation energy (Moran and Shapiro, 2007). Figure 2.6 shows the general flow chart and the exergy flow chart of a reversible fuel cell suitable for power generation if a flow of substance j at high concentration is "expanded" to a lower concentration. For the determination of transformation energy, these flows must be at reference temperature and pressure.

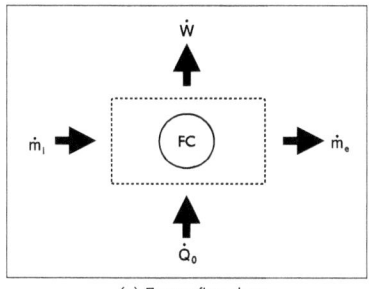
(a) Energy flow chart

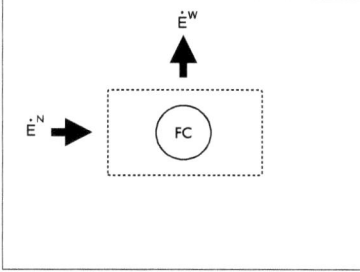
(b) Exergy flow chart

Figure 2.6: Flow charts of a reversible fuel cell process used for the derivation of nonreactive transformation energy

The energy balance of a reversible isothermal fuel cell at T_0, p_0 with the considered substance j undergoing no change of aggregate state can be expressed as:

$$0 = \dot{m}_j \cdot (h_j - h_{j,0}) - \dot{W} + \dot{Q}_{j,0}$$

The general process flow chart of the reversible fuel cell used for the evaluation of nonreactive transformation energy is similar to that of the reversible expansion process, which becomes apparent if comparing Figures 2.4 and 2.6. The nonreactive transformation energy flow $\dot{En}_{\tau,j}^N$ associated with a substance flow can therefore be defined analogously to the mechanical transformation energy flow as the net heat input flow into the fuel cell at reference temperature:

$$\dot{En}_{\tau,j}^N = \dot{m}_j \cdot (h_j - h_{j,0}) + \dot{Q}_{j,0}$$

Using the expression for heat at reference temperature from Equation 2.21 obtained for the reversible expansion process, the heat input required for a reversible fuel cell process considering only one substance is:

$$\dot{Q}_{j,0} = -\dot{m}_j \cdot T_0 \cdot (s_j - s_{j,0})$$

A nonreactive transformation energy flow can consequently be expressed as:

$$\dot{En}^N_{\tau,j} = \dot{m}_j \cdot [h_j - h_{j,0} - T_0 \cdot (s_j - s_{j,0})] \qquad (2.29)$$

The nonreactive exergy flow for a substance j at a mole fraction x_j larger than or equal to the mole fraction of the substance in the environment $x_{j,0}$ is defined as:

$$\dot{E}^N_j = \dot{m}_j \cdot [h_j - h_{j,0} - T_0 \cdot (s_j - s_{j,0})]$$

Thus, nonreactive transformability τ^N associated with substance flows at $x_j \geq x_{j,0}$ can be calculated as:

$$\tau^N_j = \frac{\dot{E}^N_j}{\dot{En}^N_{\tau,j}} = 1$$

For a substance at a mole fraction $x_j < x_{j,0}$, the higher concentration of the considered substance is found in the environment. Consequently, if a fuel cell can generate power from a concentration difference, then the high concentration source which makes it possible to generate work from interaction of the combined system, is the environment.

The derivation that leads to Equation 2.29 is also valid for substance flows at concentrations below reference concentration. Therefore, the nonreactive transformation energy can be expressed as:

$$\dot{En}^N_\tau = \dot{m}_j \cdot [h_j - h_{j,0} - T_0 \cdot (s_j - s_{j,0})] \qquad (2.30)$$

For substance flows where $x_j < x_{j,0}$, the specific entropy $s_j > s_{j,0}$ thus it follows that:

$$-T_0 \cdot (s_j - s_{j,0}) < 0$$

In analogy to specific mechanical exergy, the specific values associated with such a substance flow have a negative sign which based on Figure 2.7 can be consistently interpreted as an indicator of an opposed direction of the exergy flow to the mass flow it is associated with.

The transformability associated with a substance flow at $x_j < x_{j,0}$ can therefore be calculated as:

$$\tau^N_j = \frac{\dot{E}^N_j}{\dot{En}^N_{\tau,j}} = 1$$

The compensation heat flow that is required in the transformation energy balance is calculated based

2 The exergy-based transformability concept

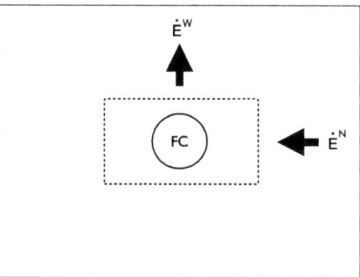

Figure 2.7: Exergy flow chart of a reversible fuel cell for substances above reference condition

on Equations 2.1 and 2.30 as:

$$\begin{aligned}
\dot{Q}_{0,j}^{*N} &= \dot{m}_j \cdot (h_j - h_{j,0}) - \dot{En}_{r,j}^{N} \\
&= \dot{m}_j \cdot \{h - h_{j,0} - [h_j - h_{j,0} - T_0 \cdot (s_j - s_{j,0})]\} \\
&= \dot{m}_j \cdot T_0 \cdot (s_j - s_{j,0})
\end{aligned}$$

If calculating total chemical exergy of fuels it is usually sensible to determine the nonreactive transformation energy and the compensation heat flow on a molar basis. This would result in the following expressions:

$$\dot{En}_{r}^{N} = \dot{n}_j \cdot [\bar{h}_j - \bar{h}_{j,0} - T_0 \cdot (\bar{s}_j - \bar{s}_{j,0})]$$

and

$$\dot{Q}_{0,j}^{*N} = \dot{n}_j \cdot T_0 \cdot (\bar{s}_j - \bar{s}_{j,0})$$

It has to be pointed out that, if evaluating the nonreactive exergy of mixtures, the increase of the mole fraction of one component always implies the decrease of the mole fraction of at least one other substance present in the considered mixture. Thus, the total nonreactive transformation energy or exergy of a mixture is always a sum of negative and positive nonreactive substance transformation energies and exergies which have to be determined for every mixture specifically. In contrast to most other forms of exergy, nonreactive exergy can in most cases not be used for power generation, due to a lack of suitable membranes (Stephan and Mayinger, 1986).

2.5 Effective thermal transformability of differences in thermal transformation energy

In exergy analysis, the exergetic efficiency is frequently defined using differences of input and exiting exergy associated with a considered mass flow. For all nonthermal transformation energies the transformability of the transformation energy equals $100\,\%$. Consequently, every difference in nonthermal transformation energy associated with a mass flow can also be associated with a transformability of $100\,\%$. To also allow a direct evaluation of differences in thermal transformation energy, it appears sensible to define *effective* thermal transformabilities τ^H as functions of differences in thermal transformation energy, which can also be referred to as effective thermal transformation energy :

$$\begin{aligned} \tau^H &= \frac{\dot{E}_i^T - \dot{E}_e^T}{\dot{En}_{\tau,i}^T - \dot{En}_{\tau,e}^T} \\ &= \frac{\dot{E}^H}{\dot{En}_{\tau}^H} \end{aligned} \qquad (2.31)$$

As a consequence, every effective thermal exergy flow $\dot{E}^H$ which is associated with a mass flow can be expressed as a product of the effective thermal transformation energy flow $\dot{En}_{\tau}^H$ associated with a massflow and the matching effective thermal transformability:

$$\dot{E}^H = \dot{En}_{\tau}^H \cdot \tau^H$$

Since thermal transformation energy is defined differently for flows at temperatures below reference temperature and for flows at temperatures above reference temperature, the effective thermal transformability is defined differently depending on the temperatures of the considered in- and outflows in relation to reference temperature. A common basis for the derivation of the effective transformability is the definition of a difference in thermal exergy or an effective thermal exergy flow [4]:

$$\dot{E}^H = \dot{m} \cdot \{[h_i - h_{T0,i} - T_0 \cdot (s_i - s_{T0,i})] - [h_e - h_{T0,e} - T_0 \cdot (s_e - s_{T0,e})]\} \qquad (2.32)$$

To obtain τ^H for a given mass flow, the difference in thermal transformation energy needs to be defined dependent on the relation of the considered temperature to the reference temperature.
For $T_i > T_0$ and $T_e > T_0$ the effective thermal transformation energy can be expressed using Equation 2.12 as:

$$\dot{En}_{\tau}^H = \dot{m} \cdot [(h_i - h_{T0,i}) - (h_e - h_{T0,e})]$$

The effective thermal transformability of the considered transformation energy difference can conse-

---

[4]This equation is valid under the assumption that $\dot{m}_i = \dot{m}_e = \dot{m}$. So that an enthalpy $\dot{H}$ flow can be expressed as: $\dot{H} = \dot{m} \cdot (h - h_0)$.

quently be calculated as:

$$\tau^H = 1 - T_0 \cdot \frac{(s_i - s_{T0,i}) - (s_e - s_{T0,e})}{(h_i - h_{T0,i}) - (h_e - h_{T0,e})}$$

In the case that no pressure change occurs between input and exit flow, the effective thermal transformation energy can be expressed in shorter form as:

$$\dot{E}n_\tau^H = \dot{m} \cdot (h_i - h_e) = \dot{Q}^H \qquad (2.33)$$

In this case, the effective thermal transformability can also be expressed as:

$$\tau^H = 1 - T_0 \cdot \frac{(s_i - s_e)}{(h_i - h_e)}$$

Using the definition of thermodynamic average temperature in Equation 2.14, the effective thermal transformability associated with an enthalpy difference of a mass flow at constant pressure and at a temperature above reference temperature can be defined in analogy to the thermal transformability associated with conductive heat flows as:

$$\tau^H = 1 - \frac{T_0}{T_a}$$

Further effective transformabilities can be defined for the other possible temperature combinations:

For $T_i < T_0 < T_e$ the effective thermal transformation energy can be expressed using Equations 2.12 and 2.17 as:

$$\dot{E}n_\tau^H = \dot{m} \cdot [-T_0 \cdot (s_i - s_{T0,i}) - (h_e - h_{T0,e})]$$

Using Equations 2.32 and 2.31, the effective thermal transformability is obtained as:

$$\tau^H = 1 - \frac{T_0 \cdot (s_e - s_{T0,e}) + (h_i - h_{T0,i})}{T_0 \cdot (s_i - s_{T0,i}) + (h_e - h_{T0,e})}$$

In the inverse case of a high temperature inflow and a low temperature exit flow $T_e < T_0 < T_i$, the effective thermal transformation energy is defined as:

$$\dot{E}n_\tau^H = \dot{m} \cdot [(h_i - h_{T0,i}) + T_0 \cdot (s_e - s_{T0,e})]$$

Consequently, the effective thermal transformability can be expressed as:

$$\tau^H = 1 - \frac{T_0 \cdot (s_i - s_{T0,i}) + (h_e - h_{T0,e})}{T_0 \cdot (s_e - s_{T0,e}) + (h_i - h_{T0,i})}$$

Finally, if $T_i < T_0$ and $T_e < T_0$, then using Equations 2.32 and 2.31 and the definition of trans-

2 The exergy-based transformability concept

formation energy for mass flows at temperatures below reference temperature in Equation 2.17, the following expressions are obtained:

$$\dot{En}_\tau^H = \dot{m} \cdot \{-T_0 \cdot (s_i - s_{T0,i}) - [-T_0 \cdot (s_e - s_{T0,e})]\}$$

and

$$\tau^H = 1 - \frac{(h_i - h_{T0,i}) - (h_e - h_{T0,e})}{T_0 \cdot [(s_i - s_{T0,i}) - (s_e - s_{T0,e})]}$$

In the case that no pressure change occurs between input and exit flows, the effective thermal transformation energy can be expressed using the definition of the thermodynamic average temperature in Equation 2.14 and Equation 2.7 as:

$$\begin{aligned} \dot{En}_\tau^H &= -\dot{m} \cdot T_0 \cdot (s_i - s_e) \\ &= -\dot{m} \cdot \frac{T_0}{T_a} \cdot (h_i - h_e) \\ &= -\frac{T_0}{T_a} \cdot \dot{Q}^H \end{aligned}$$

This is an analogous definition to the transformation energy definition found for conductive heat flows below reference temperature presented in Equation 2.8. The effective transformability of the effective thermal transformation energy can also be expressed using Equation 2.14 as:

$$\begin{aligned} \tau^H &= 1 - \frac{(h_i - h_e)}{T_0 \cdot (s_i - s_e)} \\ &= 1 - \frac{T_a}{T_0} \end{aligned}$$

The results for the effective thermal transformation energy of mass flows at constant pressures show that the effective thermal transformability is very similar to the transformability obtained for conductive heat flows. The major difference is the use of the thermodynamic mean temperature instead of the constant temperature at which conductive heat flows are considered. However, this result is only valid if the mass flows under consideration enters and exits either above or below reference temperature. For the uncommon cases, in which input and output temperatures are on different "sides" of the reference temperature, the more complicated expressions derived in this section have to be utilized.

For a discussion of the consideration of compensation heat flows in transformation energy balances and for effective thermal transformation energy flows see subsection 3.2.3 on page 54 ff.

2.6 Average transformability and total transformation energy

In practice, mass flows frequently differ in more than one intensive property from the environment. Therefore they can be associated with more than one type of transformability, e.g. thermal and

2 The exergy-based transformability concept

mechanical. If all types of specific exergy associated with the flow have the same algebraic sign [5], average transformability and total transformation energy can be used to characterize the considered flow as a whole. It is also possible to define an average transformability for multiple flows going into one direction in relation to the considered balance boundary, such as an average fuel transformability which is useful when defining transformability based ratios.

The average transformability τ_a associated with one flow or with multiple transfers having the same direction can be defined, using an expression for the total transformation energy flow associated with the fuel flow $\dot{E}n_{\tau,F}^{TO} = \sum_X \dot{E}n_{\tau,F}^{X}$ and the total exergy flow associated with the fuel flow $\dot{E}_F^{TO} = \sum_X \dot{E}_F^X$, as:

$$\tau_{a,F} = \frac{\dot{E}_F^{TO}}{\dot{E}n_{\tau,F}^{TO}} \quad (2.34)$$

In the context of the definition of average transformabilities, it is also possible to include transformation energy differences and the matching effective transformabilities into this equation.

To obtain a consistent expression for the different types of transformation energy associated with a considered mass flow, it is necessary to define a sequence of transformations. This is required since the evaluation of mechanical and chemical transformation energy assumes the possibility of heat inputs at reference temperature and since chemical exergy is evaluated for flows at reference temperature and pressure. Extending the principles for the evaluation of thermal and mechanical exergy presented by Lazzareto and Tsatsaronis (2006), a sequence for the evaluation of the different types of transformation energy, transformability and consequently exergy, can be devised as follows:

1. Evaluate thermal transformability and transformation energy at the pressure and chemical composition of the flow.

2. Evaluate mechanical transformation energy at reference temperature and the chemical composition of the flow.

3. Evaluate chemical or nonreactive transformation energy at reference temperature and pressure.

Alternatively to the calculation of the average transformability based on Equation 2.34, the average transformability can be calculated using the different types of transformability and transformation energy only. The average transformability associated with a pure substance mass flow at $T \geq T_0$, $p \geq p_0$ and $e^{CH} \geq 0$ can also be expressed as a function of thermal, mechanical and chemical transformabilities if all relevant transformation energy flows have the same direction in relation to the

---

[5] The definition of an average transformability is only sensible under the condition that all types of specific exergy (especially thermal and mechanical) have the same algebraic sign, which if considering one mass flow is usually only the case for positive exergy and transformation energy values, since thermal specific exergy cannot become negative. Without this condition average transformability values greater 100 % could result, thus violating basic characteristics of the transformability definition.

considered system:

$$\begin{aligned}
\tau_a &= \frac{\dot{E}^{TO}}{\dot{E}n_\tau^{TO}} = \frac{\dot{E}^T + \dot{E}^M + \dot{E}^{CH}}{\dot{E}n_\tau^T + \dot{E}n_\tau^M + \dot{E}n_\tau^{CH}} \\
&= \frac{\dot{E}n_\tau^T}{\dot{E}n_\tau^{TO}} \cdot \frac{\dot{E}^T}{\dot{E}n_\tau^T} + \frac{\dot{E}n_\tau^M}{\dot{E}n_\tau^{TO}} \cdot \frac{\dot{E}^M}{\dot{E}n_\tau^M} + \frac{\dot{E}n_\tau^{CH}}{\dot{E}n_\tau^{TO}} \cdot \frac{\dot{E}^{CH}}{\dot{E}n_\tau^{CH}} \\
&= \frac{\dot{E}n_\tau^T}{\dot{E}n_\tau^{TO}} \cdot \frac{\dot{E}^T}{\dot{E}n_\tau^T} + \frac{\dot{E}n_\tau^M}{\dot{E}n_\tau^{TO}} + \frac{\dot{E}n_\tau^{CH}}{\dot{E}n_\tau^{TO}} \\
&= \sum_X \frac{\dot{E}n_\tau^X}{\dot{E}n_\tau^{TO}} \cdot \tau^X
\end{aligned} \qquad (2.35)$$

Defining the transformation energy factor as:

$$f_\tau = \frac{\dot{E}n_\tau^X}{\dot{E}n_\tau^{TO}} \qquad (2.36)$$

the average transformability associated with mass flows above reference temperature and pressure can be expressed as:

$$\tau_a = \frac{\sum_X \dot{E}n_\tau^X \cdot \tau^X}{\dot{E}n_\tau^{TO}} = \sum_X f_{\tau,i} \cdot \tau_i^X$$

With this equation, the average transformability and total transformation energy can be calculated from Tables A.2 on page 130 and A.3 on page 131 without the need of calculating exergy values.

3 Evaluation of energy supply systems and thermodynamic analysis using the transformability concept

In this chapter, transformation energy efficiency and transformability ratio are developed on the basis of the transformation energy and the exergy balance as assessment ratios applicable to thermodynamic analysis and comparative assessment of thermal systems - the transformability evaluation and analysis method. Since a focus of this work lies on the application of the transformability concept to comparative assessment of energy supply systems, an instructive approach to the definition of energy system boundaries is discussed. Finally, a method for a consistent attribution of fuel to heat from combined heat and power processes is presented. The combination of this attribution method with the concept of average transformability allows the association of a characteristic quality to heat from combined heat and power, thus distinguishing it from waste heat as well as from heat generated in boiler systems.

3.1 The transformation energy balance

The basis of comparative evaluation and thermodynamic analysis using the transformability concept is the transformation energy balance. The quantitative properties on which the transformation energy concept is based are exergy and energy. Both properties can be used to draw balances. As transformation energy is a derivative property of the two properties mentioned, a balance should also be possible. Since transformation energy should, like energy, satisfy the law of conservation while having the same direction as exergy, several adaptations have to be made to fulfill these requirements. It has been mentioned in subsection 2.2.2 on page 24 ff. that the transformation energy balance is more complex than energy or exergy balances due to its characteristics that are combined from exergy (direction) and energy (law of conservation). Therefore, in the transformation energy balance, not only the transformation energy associated with the flow under consideration has to be taken into account, but also the compensation heat flows, e.g. for heat transfers at temperatures below reference temperature and mechanical transformation energy flows.

To illustrate the way in which the transformation energy concept differs from the exergy and the energy concept, flow charts of a reversible heat engine used to determine the thermal transformation energy associated with a mass flow at a temperature below reference temperature are shown in Figures 3.1 and 3.2.

3 Evaluation of energy supply systems and thermodynamic analysis using the transformability concept

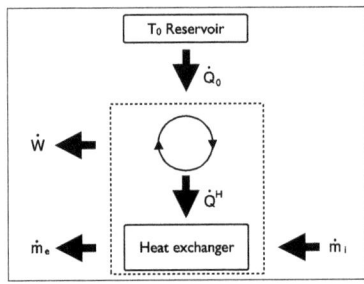
(a) Energy and mass flow chart

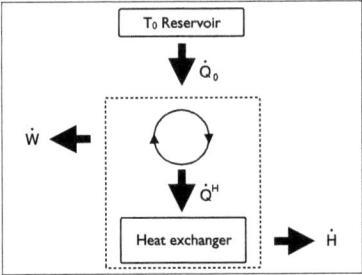
(b) Energy and enthalpy flow chart

Figure 3.1: Flow charts of a reversible heat engine process operating between reference temperature and a mass flow at a temperature below reference temperature - part 1

Figure 3.1 shows that the enthalpy flow $\dot{H}$ is opposed to the mass flow. This is caused by choosing the enthalpy reference state equal to the reference state of exergy. Since $h_i < h_0$, the enthalpy flow has a negative sign and is thus opposed to the considered mass flow. The exiting mass flow $\dot{m}_e$ is not considered an enthalpy flow since the specific enthalpy of the flow equals reference enthalpy. The heat flow $\dot{Q}^H$ causing the temperature increase of the considered mass flow $\dot{m}_i$ has a direction opposed to the enthalpy flow and must therefore have a different algebraic sign. The resulting heat flow equals the enthalpy flow and can be calculated as a function of the specific enthalpy h of the input (subscript i) and at reference conditions (subscript 0) as:

$$\dot{Q}^H = -\dot{m}_i \cdot (h_i - h_0) = -\dot{H}$$

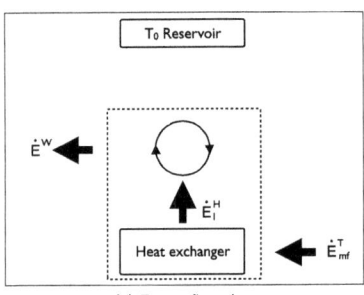
(a) Exergy flow chart

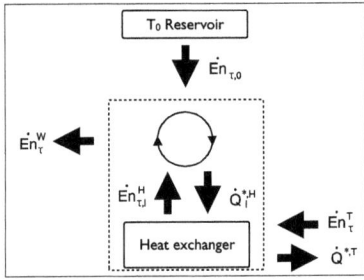
(b) Transformation energy flow chart

Figure 3.2: Flow charts of a reversible heat engine process operating between reference temperature and a mass flow at a temperature below reference temperature - part 2

Figure 3.2 shows the exergy and the transformation energy flow charts. Using the interpretation of a negative sign as an indicator of direction, the thermal exergy flow $\dot{E}^T_{mf}$, which is defined by Equation

3 Evaluation of energy supply systems and thermodynamic analysis using the transformability concept

2.13, is opposed to the enthalpy flow it is associated with but has the same direction as the mass flow $\dot{m}_i$. Similarly, according to Equation 2.5 for effective heat flows at temperatures below reference temperature, the heat flow is opposed the exergy flow it is associated with, which is why $\dot{E}_l^H$ has a different direction than $\dot{Q}^H$. Heat at reference temperature is not associated with an exergy flow, the same is valid for mass flows at reference conditions. The transformation energy flow chart is more complex as it consists not only of the relevant transformation energy flows but also of the compensation heat flows, which are necessary to fulfill the energy balance of the considered system. However, based on this balance a comprehensive method can be developed which allows the use of transformation energy and transformability as basic properties for system evaluation ratios like e.g. efficiency.

3.2 Transformability ratio and transformation energy efficiency

The goal of comparative technology assessment of energy supply systems can be defined as evaluating different technology options with regard to a specific predefined task and quantifying the differences that set the considered options apart. The comparative transformability assessment is meant to answer the question: „What technology is the most suitable and most sophisticated to perform a given supply task?"

Thermodynamic technology analysis aims at identifying improvement potential within complex systems. The major difference between comparative technology assessment and thermodynamic analysis is the balance boundary. To obtain meaningful results, it appears necessary that for comparative assessment the boundary has to be defined in such a way that the input flows are all subject to the same defining rules and the output flows are the same for all considered systems[1]. In contrast to that, the boundaries used in a thermodynamic analysis can be adapted freely in such a way that required information regarding a specific process or component can be obtained.

Despite this difference, a transformability based comparative assessment and a matching analysis method both require ratios which allow an understanding of the improvement potential. The goal of the following discussion is to find appropriate ratios in order to allow system assessment and analysis using ratios based on the transformation energy and transformability concept.

The transformation energy and transformability definitions for energy and mass transfers presented in chapter 2 lay the basis for a novel assessment and analysis method, which attempts a separate evaluation of suitability of a technology for a given supply task and its degree of sophistication.

3.2.1 Exergetic efficiency as a basis for transformability based evaluation ratios

Exergetic efficiency provides a measure of the total degree of sophistication of the considered technology option or component and is thus a useful measure for ranking systems accordingly. An exergetic efficiency below 100 % indicates that the process is not reversible, which can be caused by exergy

---

[1] If the boundary definition for the compared options does not fulfill this criterion, it is possible that the boundary definition and not the technological performance has the greatest influence on the results of a comparative evaluation.

losses and exergy destruction. The differentiation between exergy losses and exergy destruction is dependent on the choice of the system boundaries chosen for the evaluation of heat and mass losses (Tsatsaronis et al., 2007). Therefore, based on the exergy concept alone, a clear distinction of exergy losses caused by mass or energy losses from exergy destruction caused by irreversibilities that are not directly associated with mass or energy losses[2] is not possible. E.g. if all loss flows cross the boundary only after their intensive properties equal those of the environment, which can be caused by interaction of loss flow and process surroundings within the balance boundary, all exergy decreases associated with mass or energy losses are considered exergy destruction. Thus, a disadvantage of the aggregated evaluation of quality and quantity aspects by using exergy is the low transparency of the exergetic efficiency evaluation. Since the transformability concept separates quantitative aspects from qualitative aspects, a set of ratios based on these properties should allow an evaluation and analysis of the considered systems where the impact of external losses is less dependent on the system boundary at which these losses are evaluated.

Exergetic efficiency has been defined as a ratio of product and fuel by Tsatsaronis (1984) and Tsatsaronis and Winhold (1985). A suitable definition of fuel and product using transformation energy has been presented in subsection 1.1.8 on page 16. In order to ensure that the definition of an exergetic efficiency is possible for all parameter choices of common processes, it is necessary to consider the different types of exergy flows separately when defining fuel and product, i.e. thermal and mechanical exergy would be assessed instead of physical exergy. The necessity of a separate evaluation is illustrated using the example of a heat pump in appendix A3 on page 126 ff.

The fuel definition summarized by Bejan et al. (1996) and Lazzareto and Tsatsaronis (2006) also includes the possibility to subtract unwanted exergy outputs or exergy increases from the total sum of fuel exergy. An inclusion of this possibility into the fuel definition is incompatible with the transformability concept as it could result in values of average fuel transformability that are larger than unity. Therefore, within this thesis, the definition of fuel exergy is limited to sums of exergy decreases and inputs while the definition of product exergy is limited to sums of useful exergy increases and exergy outputs. In those cases where unwanted thermodynamic inefficiencies occur, such as mass or heat losses to the environment, they are not considered in the exergetic efficiency.

3.2.2 Definitions

With the introduction of the compensation heat flows, a transformation energy balance can be performed as universally as an exergy balance, thus making it possible to define a transformation energy efficiency in analogy to exergetic efficiency. The rules for defining product and fuel equal those valid for the calculation of exergetic efficiency, taking the aspects previously discussed into account. However, it is required to additionally consider the so called effective compensation heat flow to obtain universally meaningful ratios. The effective compensation heat flow is defined in subchapter 3.2.3 on page 54 ff.

The product of a process or component in terms of transformation energy is defined as the sum of:

---

[2] such as pressure losses or heat exchange in real heat exchangers

- all useful transformation energy outputs associated with mass-free energy transfers
- all increases of transformation energy between input and output associated with mass flows that are considered useful, taking into account all types of transformation energy separately
- the effective compensation heat flow if it is an output

The fuel of a process is defined analogously but with a different direction in relation to the process as the sum of:

- all transformation energy inputs associated with mass-free energy transfers
- all decreases of transformation energy between input and output associated with mass flows, taking into account all types of transformation energy separately
- the effective compensation heat flow if it is an input

The transformation energy flow associated with mass flows is a function of its temperature, pressure and composition. Thus, it is necessary to define a system boundary at which mass losses from the considered system or component are evaluated. In analogy to the definition of the balance boundary for the exergetic evaluation of conductive heat losses (Tsatsaronis et al., 2007), mass flows discharged to the environment without use are best considered to be at reference temperature, pressure and composition. They are thus not associated with transformation energy or compensation heat flows. However, frequently transformability destruction occurs within such a system boundary, which brings mass flows from the conditions at which they are really lost to reference conditions. This transformability destruction is usually compensated by additional fuel input into the process or component and can be accompanied by heat losses to the environment at reference temperature. Since these effects have an influence on transformation energy efficiency, it will nonetheless indicate improvement potential due to mass losses.

Exergetic efficiency ε [3] is defined as a function of the exergy flows $\dot{E}$ of product (subscript P) and fuel (subscript F):

$$\varepsilon = \frac{\dot{E}_P}{\dot{E}_F}$$

The product exergy flow can be expressed as a sum of all types[4] (superscript X indicates a type) of useful exergy output flows $\dot{E}^X_{e,U}$ that are associated with mass-free energy transfers and of all types of useful exergy flow increases $\Delta \dot{E}^X_{e,U} = \dot{E}^X_{e,U} - \dot{E}^X_{i,U} > 0$ that are associated with mass flows. The fuel term can be defined as a sum of all types of exergy input flows $\dot{E}_i$ associated with mass-free energy transfers and of all types of exergy flow decreases $\Delta \dot{E}_i = \dot{E}_e - \dot{E}_i < 0$ of mass flows. The symbol $\sum$ denotes a sum over all exergy flows[5] that fall into the relevant category: either input (index i) or useful exit flow (index e, U). Using these symbols, exergetic efficiency can be expressed as:

---

[3] Exergetic efficiency is used synonymously to rational transit-free exergetic efficiency. The transit definition used has been labeled transformation-oriented transit and goes back to a method presented by Sorin, Brodyansky and Valero (Riedl, 2006).

[4] It appears sensible to define exergetic efficiency on the basis of a separate evaluation of all types of exergy flows, since only in such way a applicability of this ratio to the maximum number of technologies can be ensured. See appendix A3 on page 126 for a discussion.

[5] Every type of exergy associated with a transfer is considered a separate flow.

3 Evaluation of energy supply systems and thermodynamic analysis using the transformability concept

$$\varepsilon = \frac{\sum \dot{E}_{e,U}^X + \sum \Delta \dot{E}_{e,U}^X}{\sum \dot{E}_i^X + \sum \Delta \dot{E}_i^X}$$

Analogously *transformation energy efficiency* η_τ can be defined as a function of the transformation energy flow $\dot{E}n_\tau$ associated with product and fuel flows and the effective compensation heat flow $\Delta \dot{Q}_i^*$ as [6]:

$$\begin{aligned}\eta_\tau &= \frac{\dot{E}n_{\tau,P}}{\dot{E}n_{\tau,F}} \\ &= \frac{\sum \dot{E}n_{\tau,e,U}^X + \sum \Delta \dot{E}n_{\tau,e,U}^X}{\sum \dot{E}n_{\tau,i}^X + \sum \Delta \dot{E}n_{\tau,i}^X + \Delta \dot{Q}_i^*}\end{aligned}$$

Since the product of transformation energy and transformability associated with a given flow equals the exergy of this flow, it appears straightforward to add another specific characteristic to the definition of the transformation energy efficiency, a defined relation to exergetic efficiency ε with the factor ξ:

$$\varepsilon = \xi \cdot \eta_\tau$$

Transposing this equation and using Equation 2.35, ξ can be expressed as a ratio of average transformabilities τ_a of product and fuel:

$$\begin{aligned}\xi &= \frac{\varepsilon}{\eta_\tau} \\ &= \frac{\dot{E}_P}{\dot{E}_F} \cdot \frac{\dot{E}n_{\tau,F}}{\dot{E}n_{\tau,P}} \\ &= \frac{\frac{\dot{E}_P}{\dot{E}n_{\tau,P}}}{\frac{\dot{E}_F}{\dot{E}n_{\tau,F}}} \\ &= \frac{\tau_{a,P}}{\tau_{a,F}}\end{aligned}$$

ξ, the relation of exergetic efficiency to transformation energy efficiency, is thus a direct measure of the match between the average quality of the useful product in relation to the average quality provided. Thus, it can be labeled *transformability ratio*, as it is the ratio of the average product transformability to the average fuel transformability. It is not an efficiency since transformability is a measure of quality that cannot be balanced on its own. The transformability ratio can also be calculated directly without having to calculate exergetic efficiency on the basis of the definitions of transformation energy, transformability, transformation energy differences and the average transformability concept presented in the previous chapter.

---

[6]In this equation it is assumed that the effective compensation heat has a positive sign, indicating a heat transfer from the environment into the process. The calculation of effective compensation heat flows is discussed in subsection 3.2.3 on the following page.

3 Evaluation of energy supply systems and thermodynamic analysis using the transformability concept

For $\Delta \dot{Q}_i^* < 0$, the effective compensation heat flow is an output of the considered system. Since the compensation heat flows, on which the calculation of the effective compensation heat flow is based, are necessarily associated with certain transformation energy flows relevant in product or fuel definition, the effective compensation heat flow should be considered a useful output, since it is required [7]. This approach leads to the following expression of transformation energy efficiency for negative effective compensation heat flows:

$$\eta_\tau = \frac{\sum \dot{En}_{\tau,U}^X + \sum \Delta \dot{En}_{\tau,U}^X - \Delta \dot{Q}_i^*}{\sum \dot{En}_{\tau,i}^X + \sum \Delta \dot{En}_{\tau,i}^X}$$

If $\Delta \dot{Q}_i^* < 0$, the effective compensation heat flow becomes a part of the definition of the average product transformability instead of the average fuel transformability when calculating ξ.

The definitions of transformation energy efficiency and transformability ratio can be expressed as follows:

The transformation energy efficiency is the ratio of product transformation energy to fuel transformation energy of the considered process.

To define the ratio consistently, the different types of transformation energy have to be considered as separate flows. The fuel transformation energy is defined as the sum of all transformation energy inputs into the considered system, which are associated with mass-free energy transfers, plus all decreases of transformation energy between input and exit, which are associated with mass flows, plus the effective compensation heat flow if it is an input. The product on the other hand is defined as the sum of all useful transformation energy outputs, which are associated with mass-free energy transfers, plus all useful increases of transformation energy between input and output, associated with mass flows, plus the effective compensation heat flow if it is an output.

The transformability ratio is the ratio of the average product transformability of the total product transformation energy and the average fuel transformability of the total fuel transformation energy. It can also be calculated from the ratio of exergetic efficiency to transformation energy efficiency if both ratios are defined following similar rules for product and fuel definition.

3.2.3 Considering compensation heat flows in transformation energy efficiency

Compensation heat flows are required to draw transformation energy balances. They are included into the transformation energy balance since transformation energy is a function of the combined system of reference environment and flow under consideration, while the energy balance is fulfilled only for the energy directly transported by the relevant input and exit flows, which is independent of the reference environment. The compensation heat flows can in general be considered as a measure of the energetic

---

[7] Not considering the effective compensation heat flow in the numerator of η_τ could result in transformability ratios larger than 100 %, thus violating the implicit reference for all transformability ratios of 100 %. Additionally, the inclusion of a negative effective compensation heat flow in the denominator (the fuel term) could result in an average input transformability greater than 100 %, which would contradict the basic assumptions of the transformability concept.

3 Evaluation of energy supply systems and thermodynamic analysis using the transformability concept

influence of the environment on the energy input required to transform the considered exergy flows into work. Since an efficiency must always have values between 0 and 100 %, the compensation heat flows have to be introduced somehow into the definition of transformation energy efficiency. Furthermore, in order to avoid transformability ratios that are greater 100 %, it has been found to be sensible to take ideally required heat flows at reference temperature into consideration, which are needed for the reversible operation of some processes. A sensible way to deal with these two types of heat flows at reference temperature is to consider the sum of all compensation heat flows $\dot{Q}^*$ and the ideally required heat transfers $\dot{Q}_0^{id}$ at reference temperature[8] in a property labelled *effective compensation heat* flow, which is defined as:

$$\Delta \dot{Q}_i^* = \sum \dot{Q}_i^* - \sum \dot{Q}_e^* + \sum \dot{Q}_{0,i}^{id} - \sum \dot{Q}_{0,e}^{id} \qquad (3.1)$$

The sum of the ideally required heat transfers from the environment $\sum \dot{Q}_{0,i}^{id}$ summarizes all heat flows from the environment that are minimally necessary for process operation e.g. a heat flow from the environment into a reversible isothermal expansion process[9]. Similarly, the sum of all ideally required heat transfers to the environment $\sum \dot{Q}_{0,e}^{id}$ stands for all heat flows that have to be discharged to the environment in order to operate a considered process, e.g a heat flow discharged by a reversible isothermal compressor to allow isothermal operation. The ideally required heat transfers are technology independent and can be obtained by evaluating the energetically ideal processes at the considered process parameters. As a consequence of the consideration of the ideally required heat flows, only losses that exceed the minimal loss of heat to the environment are decreasing the transformation energy efficiency. Since the heat discharge of a considered process can occur at nonoptimal parameters, e.g. a heat engine discharging heat at temperature above reference temperature, transformability destruction can occur even in these energetically ideal processes. This yields a better characterization of the real improvement potential of the process at the expense of increasing the complexity of the evaluation since reversible operation parameters have to be calculated and considered additionally. The calculation of ideally required heat flows for some common processes is discussed in greater detail in appendix A6 on page 133 ff.

The effective compensation heat flow has to be considered like any other transformation energy flow in the transformation energy efficiency, but does not need to be considered in exergetic efficiency since the associated transformability is zero. However, in case the effective compensation heat flow is an output of the considered system, it has to be taken into account as a useful flow in the transformation energy efficiency, since all heat flows comprising this difference are unavoidable for process operation and the transformability balance.

Although a consistent way of dealing with compensation heat flows and ideal heat exchanges with the environment has been found, many other ways of dealing with these heat flows appear possible. Since

---

[8] Heat transfers at reference temperature that are necessary even for reversible operation of some processes.
[9] It is important to consider the flow chart before defining the necessary ideal heat flows. E.g. if only a heat engine is considered and heat at a high temperature is the input flow, work can only be generated by discharging heat to the environment. Thus, an ideally required heat flow is included in the calculation of the effective compensation heat flow. On the other hand, if a combustible fuel is the input into a heat engine process, work can also be generated without heat discharge e.g. by an ideal fuel cell, thus no ideally required heat flow is considered in this case.

the way in which compensation heat flows and heat exchanges with the environment are dealt with significantly influences the values of transformability ratio and transformation energy efficiency, the alternatives to the chosen approach will be discussed briefly below. These alternative approaches are also consistent with the transformability concept, but would lead to transformability evaluation and analysis methods that are considered less advantageous than the one presented in this dissertation.

Generally, a consideration of compensation heat flows and ideally required heat flows from and to the environment is necessary to obtain consistent evaluation ratios. Only by considering these heat flows, transformation energy efficiency and transformability ratio can be defined in such a way that for all applications both values do not exceed $100\,\%$. The limitation to $100\,\%$ ensures that the difference to ideal operation is clearly assessable on first glance. Consequently, it was necessary to find a way of considering the compensation heat flows and ideally required heat flows to or from the environment that leads to universally consistent and meaningful results.

For example, if compensation heat flow inputs were considered separately from compensation heat flow outputs, significant amounts of heat at reference temperature would influence the sum of fuel transformation energy, thus decreasing the influence of transformation energy flows with a nonzero transformability on the considered assessment ratio. This explains why the consideration of an effective compensation heat flow appears to be more sensible than the separate consideration of in- and outputs.

Another idea that has been also been rejected, is to avoid consideration of the effective compensation heat flow in the numerator of the transformation energy efficiency. This approach would result in a decrease of transformation energy efficiency caused by losses of unavoidable heat flows. A decrease of an efficiency that should be considered a measure of thermal sophistication by unavoidable losses appears not to be sensible. Additionally, for some applications such as a reversible refrigeration machine operating at temperatures below reference temperature, a transformability ratio larger $100\,\%$ could result since exergetic efficiency could exceed transformation energy efficiency.

Furthermore, it was attempted to evaluate thermal and mechanical compensation heat flows separately. The idea behind this separation was, to only consider those compensation heat flows as useful which are of the same type as the desired product flow, consequently decreasing for example the influence of mechanical compensation heat flows on heating applications. However, this approach would lead to transformability ratios above $100\,\%$ for some applications such as a reversible adiabatic expander and was therefore not followed through.

In addition to the options available when dealing with the compensation heat flow, the way of dealing with ideally required heat inputs and discharges to or from the environment has also been chosen from a set of alternatives. In principle, at least two other approaches regarding the consideration of ideally required heat flows could have been followed. On one hand the process-specific heat flows that are necessary even in reversible operation could have been neglected. This could result in transformability ratios larger than $100\,\%$ for processes like the reversible heat engine, which has an ideal exergetic efficiency of $100\,\%$ and an ideal energy efficiency equal to the Carnot efficiency. On the other hand the real nonuseful heat exchanges with the environment could have been considered instead of the ideally required ones. This would result in an influence of all negative effects of avoidable losses in the transformability ratio alone, leaving the transformation energy efficiency at $100\,\%$ even for

3 Evaluation of energy supply systems and thermodynamic analysis using the transformability concept

non-energetically ideal processes. Choosing from these alternatives, it appears consequently most sensible to follow the outlined approach of considering only the necessary interaction of process and environment by using the sums of the ideally required heat transfers $\sum \dot{Q}_{0,i}^{id}$ and $\sum \dot{Q}_{0,e}^{id}$.

3.2.4 Interpretation of transformation energy efficiency and transformability ratio

With transformation energy efficiency and transformability ratio, two novel assessment parameters have been defined that can complement and help to explain exergetic efficiency. While exergetic efficiency indicates the total degree of sophistication that is decreased by exergy destruction and exergy loss, transformation energy efficiency is only influenced by transformation energy losses to the surroundings that exceed unavoidable losses, such as ideally required heat losses from thermally driven heat engines.

Transformation energy efficiency thus becomes an indicator of the lack of avoidable external losses of the process. Measures that are required to increase transformation energy efficiency can frequently be applied without changing process parameters or its structure, by such means as recovering waste heat or improving thermal isolation. As a consequence, *transformation energy efficiency can be considered as a degree of external sophistication*.

Complementing this evaluation ratio, the transformability ratio provides insight into the degree of process suitability, indicating how well the average quality of a given set of fuel flows is used in the considered process[10]. It is independent of the quantitative relation of the product to the fuel flows and only dependent on the quantitative relations between the different summands in the fuel term or the product term. The comparison of transformability ratios for a given set of fuels therefore allows to identify most suited process for the use of these fuel flows. If the product is the same for all considered processes, the transformability ratio indicates what set of fuel inputs is best suited to generate it. The transformability ratio thus becomes an indicator for the process suitability in regard to a required product.

Despite its uses as a means to quantify the degree of quality match between supply and demand, the transformability ratio of the process does not allow to quantify the general suitability of a process type for the considered supply task, since the transformability ratio is decreased by most effects that decrease exergetic efficiency. However, if the transformation energy efficiency is significantly higher than the transformability ratio, e.g. $\eta_\tau = 70\,\%$ and $\xi = 10\,\%$, the largest part of theoretical optimization potential for the given process can be achieved by changing process parameters or design. If it is of interest to quantify the general suitability of a process design or a component regarding a specific product, the maximum transformability ratio can be used for this purpose.

To obtain a value for the maximum transformability ratio, the process mass and energy flows have to be calculated starting with the product flow assuming the absence of friction and of avoidable pressure and heat losses[11]. The transformabilities of all fuel flows remain constant and the direction of fuel and product flows do not change. Such processes have always a transformation energy efficiency

---
[10] The major question answered is: How well is the considered process suited to provide the product flow(s)?
[11] e.g. the ideally required heat discharge of a reversible heat engine is considered an unavoidable heat loss.

3 Evaluation of energy supply systems and thermodynamic analysis using the transformability concept

of 100 % and are additionally free of avoidable pressure drops. The maximum transformability ratio then provides a measure of the suitability of the process type, which can only be further improved by changing process design, process parameters such as temperatures and pressures or by replacing the considered system with a more suitable one. The maximum transformability ratio is identical with the maximum exergetic efficiency that is achievable with the considered process design or component. An example of a process with a low maximum transformability ratio is the boiler in which the use of a combustible fuel to generate heat by combustion always leads to significant transformability destruction [12].

Transformability ratio and transformation energy efficiency allow to quantify exergetic process performance on two complementary scales . With such a more differentiated assessment, an exergy-based process evaluation can be enhanced by shifting the focus from an evaluation of the total process sophistication (by using exergetic efficiency), to its external sophistication and to the answer to the question, how well a process uses the specific potential of its fuel, which allows to assess, how well the considered system is suited for a given supply task. Additionally, using the maximum transformability ratio, the degree of suitability of the process design can be assessed. Since this ratio equals the maximum exergetic efficiency achievable with a considered process design the introduction of the maximum transformability ratio solely allows to communicate the result in an novel way but brings no significant further benefit over the use of a maximum exergetic efficiency achievable with a considered design. In general, transformability ratio and transformation energy efficiency are not intended to replace but rather to complement exergetic efficiency to gain a deeper understanding of the evaluation results.

The interpretation of transformation energy efficiency as a degree of external sophistication and of the transformability ratio as a degree of process suitability allows to use these ratios to improve communication of the results of thermodynamic analysis, especially to people not familiar with the exergy concept. Transformation energy efficiency alludes to the commonly known energy efficiency and is in fact nearly[13] identical with it for thermal processes operating above reference temperature. As it is more universally applicable, it allows a sort of energy efficiency assessment of almost all processes and components. Adding to this universal energy efficiency, the transformability ratio can be communicated as a ratio of energy quality, thus allowing to easier understand that a fundamental engineering challenge apart from decreasing losses to the environment lies in the choice and design of processes which match product and fuel quality. Finally, the indication of the external sophistication in conjunction with the degree of process suitability can be used to identify the more significant area of improvement potential of a process.

---

[12]Transformability destruction within a process can be defined as:

$$\tau_D = \tau_{a,P} - \tau_{a,F}$$

[13]The small difference of chemical exergy from the higher heating value of a fuel accounts for the difference of an accurately calculated transformation energy efficiency from an ordinary energy efficiency.

3 Evaluation of energy supply systems and thermodynamic analysis using the transformability concept

3.3 Setting evaluation boundaries for a cross-technology comparison of supply systems

In section 1.3 on page 19 ff. it has been explained that one of the most common approaches to system evaluation is the evaluation of the steady-state operation of a supply system. It allows a comparative assessment of the operation, thus neglecting fuel mining and transportation as well as the requirements for construction and deconstruction. Although the obtained results are not as comprehensive as those of a cumulative approach, they allow the assessment of the central task of the energy system, thus laying the basis for a more extensive analysis. The evaluation of steady-state operation requires only consumption and product data and can usually be calculated for most supply systems using freely available data. Thus, for demonstration of the transformability assessment method it appears sufficient to perform a supply system comparison based on the evaluation of steady-state operation. A short overview on the definition of the energy supply systems as discussed below can be found in appendix A9 ff.

The primary aim of heating and cooling is to keep a target temperature constant[14] within a considered supply target volume. To achieve this, unwanted heat exchanges with the environment have usually to be compensated for by a supply system for heating or cooling. Such supply systems can be based on a variety of different processes, which can be assessed using various evaluation parameters in order to identify best and good practice solutions. A thermodynamic assessment allows a technological evaluation of the considered energy systems. Such an assessment makes an objective scientific evaluation of the considered supply technologies possible and is central to all multidimensional assessment methods for energy systems.

An exergy-based assessment method can provide the most comprehensive thermodynamic evaluation of such supply systems as it allows the consideration of quantitative and qualitative aspects of all types of transfers. Comparative transformability assessment allows to evaluate quantitative and qualitative effects separately. This is promising to increase the transparency and the communicability of the evaluation results and widens the options for ranking the systems, since with transformation energy efficiency and transformability ratio two efficiency-like assessment ratios are available on which such a ranking can be based.

To allow comparison of a variety of supply system alternatives, it is important to find a definition of system boundaries at which input and output are evaluated, so that the assessment results are only influenced by process characteristics and not by inconsistent boundary definitions.

3.3.1 Specifying the supply target

Supply systems are built to meet a demand. Therefore, every supply system is connected to a supply target. This target is characterized by a definition of the total demand that the supply system has to provide. The task of heat and cold supply systems can be expressed as providing a specific amount of transformation energy associated with a conductive heat flow to the supply target (e.g. a building).

---

[14]This is an abstraction of the supply task, since in reality the target temperature has to be kept within a temperature range.

3 Evaluation of energy supply systems and thermodynamic analysis using the transformability concept

Thus, only the total demand and not the causes for this demand are important for the definition of the supply target. It is therefore sufficient to define the demand of the supply target in terms of required transformation energy and the target temperature, which has to be held up versus the average outside temperature.

When defining the transformation energy demand, the advantage of the greater transparency that the transformability concept can provide becomes apparent. It allows to distinguish supply targets with high transformability and a low transformation energy demand from those with a low transformability and a high transformation energy demand.

To simplify the comparison, it is assumed that the energy demand for heating or cooling of the considered building is independent of the outside temperature. This can be achieved in practice by adapting the insulation of the building accordingly. Thus, heating supply systems can be compared for different reference temperatures, since independently of reference temperature they provide the same energetic supply. However, the influence of the reference temperature on the transformation energy associated with heat flows at a temperature below reference temperature results in changing transformation energy demands for different reference temperatures if the cooling demand is assumed to be constant.

The target supply temperature of heating and cooling is set to a room temperature of $295\,K$. Since this temperature has to be kept constant, this temperature and not an average temperature has to be considered. Since performance characteristics and costs of supply technologies depend on size, it appears sensible to define a total energy demand for the supply target in addition to the target supply temperature, so that only supply systems are compared which are able to provide this supply.

3.3.2 Defining cross-comparable subsystems

On earth all available primary energy has been generated from solar energy or is a direct use of energy from gravitational or nuclear forces. The most basic form of universally applicable boundaries at which energy inflows could be evaluated is the evaluation of all input flows at the time of their origination from one of these sources. Obviously, this is impractical as synthesis paths for all fossil fuels would have to be assumed in order to calculate the use of solar and gravitational energy required for their generation. To avoid this complication, it appears sensible to define the energy supply system boundaries according to universally applicable rules, thus providing an alternative common basis for boundary definition.

Any universal rule used for such definition of energy system boundaries could in principle lead to the neglection of vital parts of considered supply systems. To circumvent this problem, it seems to be reasonable to divide the considered supply systems into parts that can be compared among different technologies and parts which, due to their high degree of technology specificity, can only be compared within a given technology group. For the cause of simplicity, the universally comparable subsystems will be referred to as "cross-comparable" while the technology specific subsystems that cannot be included into cross-reference comparison can be labelled as "technology-specific".

The basic idea for the definition of a system boundary at which energy transfers coming into the system are evaluated is to define it in such a way that all incoming transformation energy flows are similar in at least one general aspect that is important with regard to the goal of the comparison. It appears sensible to first exclude all forms of transformation energy from the evaluation that are technologically not yet usable, such as fusion transformation energy. In the subgroup of technologically usable primary energies it appears reasonable to differentiate between storable and nonstorable primary energy. Storable primary energy forms can be stored directly while some renewable primary energy forms such as wind or solar radiation have to be converted into other forms of energy to allow storage.

Storable energy is required to ensure steady-state operation, since nonstorable primary energy cannot always be provided to a supply system on demand. Additionally, nonstorable transformation energy is either used or lost directly to the environment. Application of the storability criterion to solar thermal heat supply systems would result in splitting them into a cross-comparable subsystem, into which hot water from the solarthermal collector enters and a technology-specific subsystem, which would allow the comparison of different types of solar collectors.

Consequently, the application of the storability criterion has the effect that it reinstates comparability of nonstorable energy forms with combustible resources. Most of these have originally required solar energy to build up (e.g. gas, coal, biomass). The transformation energy efficiency from solar radiation to fuel is usually neglected if theses resources are evaluated as primary energy. It appears therefore logically sound to use the storability criterion as the major criterion for the definition of the cross-comparable supply system boundaries on the fuel side.

Based on the storability criterion, the fuel flows of the energy supply system are evaluated at the primary energy input into the conversion process or just after the conversion of a nonstorable primary energy into a storable energy form. Transportation networks are included into the cross-comparable subsystem while transportation by vehicles is excluded.

The minimum requirements for the definition of cross-comparable subsystems for heating and cooling are:

1. A common supply target, which is defined by an amount of required energy or transformation energy and the target temperature. The use of the minimally required product instead of the provided product in the efficiency definitions allows to include the identification of improvement potentials resulting from a nonoptimal choice of supply temperatures in the house, thus ensuring the overall comparability[15].

2. Knowledge of the first energy conversion process that allows to evaluate input flows of storable primary energy into the energy supply system. Due to the storability criterion, the energy supply system includes the first conversion process for storable primary energy forms while in case of nonstorable primary energy the energy transformer from nonstorable to storable energy is separately evaluated in a technology-specific subsystem.

Despite this consistent approach to boundary definition, the only thermal supply systems using nonstorable primary energy directly, the solar thermal heating systems, will not be evaluated. Since,

---

[15] If the real average supply temperatures of the in-house heating system were used as a basis for the definition of exergetic efficiency, instead of the minimally required temperatures, the evaluation ratios of the supply system could improve with increasing temperatures of the in-house heating system. High temperatures of the in-house heating system are no characteristic of better performance if supply systems are compared.

3 Evaluation of energy supply systems and thermodynamic analysis using the transformability concept

as has been briefly explained in subchapter 1.1.7 on page 15, a transformability definition suitable for solar radiation has yet to be developed. However, the evaluation of the cross-comparable subsystem of a solar thermal heating system equals in principle the evaluation of heat supply by waste heat or heat from deep geothermal sources, since the relevant transformability associated with heat from those sources is simply the effective thermal transformability associated with the heat transfer from the provided hot fluid. See appendix A10.2 on page 146 ff. for an example of a geothermal heating system.

Additional evaluation rules are required for technologies that generate more than one useful output, such as combined heat and power. The basic approach chosen for the identification of such rules is the derivation of attribution factors, which allow an attribution of defined shares of the input fuel to the products generated. This allows a separation of such multifunctional systems into separate single purpose systems, where the system providing thermal energy is the cross-comparable one. This aspect is discussed in depth in section 3.4 ff.

As a consequence of the discussed separation of energy supply systems into subsystems, technology-specific subsystems that are not considered in the cross technology comparison are transformation technologies that generate storable from nonstorable forms of energy such as solar panels and wind generators and subsystems of cogeneration plants that generate product streams that are not required by the defined supply target.

3.3.3 Evaluating energy supply technologies

Once a supply target is defined, various supply systems can be identified that can fulfill the supply task. Every system is then specified by the definition of the balance boundaries. To identify all relevant flows, an energy balance is performed for the cross-comparable subsystems. After that, using transformation energy and exergy balances and efficiencies, the transformability ratio can be calculated. Although a one-dimensional assessment based on exergetic efficiency can still be performed, the comparative transformability evaluation allows an extended two-dimensional and therefore more transparent assessment of the considered technologies. A graphical example of such a two dimensional evaluation of supply systems at different reference temperatures is discussed in section 4.3 on page 72 ff.

3.4 Evaluation of the heat output from combined heat and power plants

Usually, it is assumed that „When comparing combined heat and power (CHP) processes with the separate generation of electricity and heat, the difference can be expressed in terms of the energy saved when choosing CHP" (Nesheim and Ertesvag, 2007). The savings that are achieved by CHP can be attributed either to heat or to electricity or partially to both products. The reasoning behind the attribution of the fuel savings to electricity is, that CHP plants usually operate according to the heat demand. The reasoning behind the attribution of the fuel savings to the generated heat is that all combustion based power plants generate waste heat, which can as well be used instead of being discharged into the environment. In addition to the question, to what product the savings from CHP

3 Evaluation of energy supply systems and thermodynamic analysis using the transformability concept

are attributed, usually reference technologies are needed to calculate the savings achieved by CHP in comparison with a separated production (Nesheim and Ertesvag, 2007). Concluding from this state of the art, it appears problematic to attribute a share of fuel to heat generated from a combined heat and power (CHP) process.

Rosen (2008a) simplifies allocation methods for carbon dioxide emissions originally presented by Phylipsen et al. (1998) by introducing an allocation factor that allows to assess the relative amount of the allocated carbon dioxide to the cogenerated products. Although the allocation is related to carbon dioxide emissions, the allocation of emissions to cogenerated products based on exergy provides an interesting starting point for the development of a fuel allocation scheme that is consistent with the exergy and thus with the transformability assessment method. Adapting the nomenclature, the carbon dioxide allocation factor f_{aCO_2} can be expressed as a function of the effective thermal exergy flow $\dot{E}_e^H$:

$$f_{aCO_2} = \frac{\dot{E}_e^H}{\dot{W} + \dot{E}_e^H} \tag{3.2}$$

It appears sensible to investigate whether this approach is suitable for fuel attribution to heat from combined heat and power.

3.4.1 Derivation of the attribution of a fuel share to heat from CHP processes

All heat engines, which are the basis of most CHP plants, underlie theoretical limitations due to the Carnot efficiency. This implies that all heat engine processes above reference temperature must discharge waste heat, minimally at reference temperature. It is obvious that heat at reference temperature discharged from a power plant should not be attributed any exergy loss or destruction since it is an unavoidable byproduct of thermal power generation. Also, all exergy loss and destruction must be attributed to heat if no electricity is generated and the cogeneration plant is functioning as a large boiler. These boundary conditions are fulfilled if exergy destruction and loss flows associated with the heat output $\dot{E}_{D+L}^H$ relates to the total exergy destruction and loss flows $\dot{E}_{D+L}^{TO}$ of the process like the effective thermal exergy flow associated with the thermal product $\dot{E}_P^H$ relates to the total exergy product flow $\dot{E}_{P,CHP}^{TO}$. This relation can be expressed as:

$$\frac{\dot{E}_P^H}{\dot{E}_{P,CHP}^{TO}} = \frac{\dot{E}_{D+L}^H}{\dot{E}_{D+L}^{TO}} \tag{3.3}$$

To find a definition for the fuel attributed to heat, Equation 3.3 can be transformed into:

$$\dot{E}_P^H \cdot \dot{E}_{D+L}^{TO} = \dot{E}_{D+L}^H \cdot \dot{E}_{P,CHP}^{TO}$$

With the addition of the product $\dot{E}_P^H \cdot \dot{E}_{P,CHP}^{TO}$ on both sides this equation can be expressed as:

$$\dot{E}_P^H \cdot \dot{E}_{P,CHP}^{TO} + \dot{E}_P^H \cdot \dot{E}_{D+L}^{TO} = \dot{E}_{D+L}^H \cdot \dot{E}_{P,CHP}^{TO} + \dot{E}_P^H \cdot \dot{E}_{P,CHP}^{TO}$$

which equals:

$$\dot{E}_P^H \cdot \left(\dot{E}_{P,CHP}^{TO} + \dot{E}_{D+L}^{TO} \right) = \dot{E}_{P,CHP}^{TO} \cdot \left(\dot{E}_{D+L}^H + \dot{E}_P^H \right) \tag{3.4}$$

The following two equations define the total exergy input associated with the fuel $\dot{E}_F$ and the exergy flow associated with the fuel attributed to heat $\dot{E}_{aF}^H$ as:

$$\dot{E}_F = \dot{E}_{P,CHP}^{TO} + \dot{E}_{D+L}^{TO}$$

$$\dot{E}_{aF}^H = \dot{E}_P^H + \dot{E}_{D+L}^H$$

Using these definitions with Equation 3.4 the following expression results:

$$\dot{E}_P^H \cdot \dot{E}_F^{TO} = \dot{E}_{P,CHP}^{TO} \cdot \dot{E}_{aF}^H$$

This equation allows the definition of an effective thermal fuel attribution factor f_{aF}^H as:

$$\frac{\dot{E}_P^H}{\dot{E}_{P,CHP}^{TO}} = \frac{\dot{E}_{aF}^H}{\dot{E}_F} = f_{aF}^H \tag{3.5}$$

This definition of the fuel attribution factor essentially equals the definition of the exergy-based carbon dioxide allocation factor presented in Equation 3.2, thus proving that the exergy-based attribution of fuel has a sound logical foundation. Recently, Dittman et al. (2009) have published a paper in which one of the two suggested options for ecological attribution of fuel to heat from combined heat and power equals the one presented here.

In addition to the exergy-based fuel attribution presented here, other attribution methods that are based on exergy or exergoeconomic analysis are available (Erlach et al., 2001; Tsatsaronis et al., 2007). However, these methods require detailed knowledge of the considered process, which is usually not sufficiently available. Therefore these methods will not be used in this work for the comparative evaluation of energy systems.

For use with the comparative transformability evaluation, Equation 3.5 can be expressed in terms of average transformability and transformation energy efficiency. Since combined heat and power processes operate above reference temperature and it is assumed that chemical transformation energy of the combustible fuels (subscript cF) is equivalent to its higher heating value HHV, the effective thermal fuel attribution factor can be expressed as a function of effective thermal transformability τ^H of the effective heat flow $\dot{Q}_{CHP}^H$ from CHP and the generated electrical work flow $\dot{W}$. With the

3 Evaluation of energy supply systems and thermodynamic analysis using the transformability concept

introduction of the exergy flow associated with the total combustible fuel input $\dot{E}_{cF}^{TO}$ into the equation, the effective thermal fuel attribution factor can be expressed as a function of thermal efficiency η^T and electrical efficiency η^{EL} [16].

$$
\begin{aligned}
f_{aF}^H = \frac{\dot{E}_{aF}^H}{\dot{E}_{cF}} &= \frac{\dot{E}^H}{\dot{E}_P^{TO}} \\
&= \frac{\tau^H \cdot \dot{Q}_{CHP}^H}{\tau^H \cdot \dot{Q}_{CHP}^H + \dot{W}} \\
&= \frac{\tau^H \cdot \frac{\dot{Q}_{CHP}^H}{\dot{E}_{cF}^{TO}}}{\tau^H \cdot \frac{\dot{Q}_{CHP}^H}{\dot{E}_{cF}^{TO}} + \frac{\dot{W}}{\dot{E}_{cF}^{TO}}} \quad (3.6) \\
&= \frac{\tau^H \cdot \eta^T}{\tau^H \cdot \eta^T + \eta^{EL}} \quad (3.7)
\end{aligned}
$$

Equation 3.7 shows that the electrical efficiency of the CHP plant has a very significant influence on the fuel attribution to heat. Thus, using the exergy-based method of fuel attribution it becomes obvious that the beneficial effect of combined heat and power increases directly with increasing electrical efficiency of the CHP plant. It is therefore recommendable to promote only combined heat and power processes with a high electrical efficiency instead of CHP processes in general, since only such processes have a sufficiently high impact on fuel attribution to justify additional investments. An example of the application of this method to the evaluation of a CHP plant can be found in appendix A10.4 on page 149 ff.

---

[16] To limit imprecision, it is important to consider electrical and thermal efficiencies which are related to higher heating value flows instead of ratios which are related to lower heating value flows. Additionally, for real processes, it is recommended to consider the heat flow form the CHP process at the temperature at the hot side of the heat exchanger which heats the district heating water Dittman et al. (2009) instead of considering the effective heat flow at the thermodynamic average temperature of forward and return flow of the district heating water.

4 Application of the transformability evaluation and analysis method

In this chapter the results of the application of the transformability assessment and analysis method to various examples are discussed. First, a comparative evaluation of thermal energy supply systems is performed, which shows that the essential problem of most considered supply systems is the insufficient match between the average input and the required transformability. Supply systems that use mainly nonthermal energy to supply thermal energy are evaluated worst in the overall comparison. The reference state influences transformation energy efficiencies of the supply systems only weakly while having a high impact on transformability ratio. The reference state therefore has to be considered a major influential factor, when assessing the quality associated with thermal energy.

In a second step, the transformability analysis method is used for thermodynamic analysis of processes. It becomes clear that a transformation energy efficiency below $100\,\%$ always implies that transformation energy losses are present. Pressure losses that are not compensated by additional fuel input influence only the transformability ratio directly. The analysis is extended to a vapor-compression cascade refrigeration machine, which is operating above and below reference temperatures. The analysis of this example indicates that transformability analysis is consistently applicable to such cross reference parameter processes. Its central benefit over an analysis using only exergetic efficiency is the clear identification of transformability destruction as the major problem for optimization of thermodynamic processes.

Additionally, a short outlook on a graphical evaluation tool for energy supply scenarios is presented, which has been termed ExergyFingerprint. It has been developed at Fraunhofer UMSICHT on the basis of the transformability concept. The graphical assessment can simplify the understanding of exergy and help to discuss improvement potentials on the scale of transformability and transformation energy with people not professionally occupied with thermodynamics.

Finally, the transformability concept is found promising to be helpful in the definition of the terms "LowEx" and "LowEx-ready".

4.1 Assessment of energy supply systems

A consistent approach to energy supply system assessment has been developed in chapter 3. In short, it includes the definition of a supply target, the identification of the input flows of storable primary energy or storable secondary energy and a method for the attribution of a fuel share to heat from combined heat and power processes.

The supply systems are compared on the basis of a supply target which is defined by an energy demand and a target temperature at which the supply target should be kept. The balance boundaries on the fuel side are set in such a way that for the considered examples all inputs are primary energy flows. For the examples discussed in this dissertation, the transportation and mining energy required to provide primary energy to the energy supply system have been neglected since all data is only exemplary and not related to specific processes.

4 Application of the transformability evaluation and analysis method

Since the output flow of the supply systems is the same for all energy supply systems, the efficiencies are determined solely by the input flow. The basic assumption for a generalized comparison is the evaluation of all supply systems at steady-state using averaged parameters for energy flows and temperatures. Further simplifications are the consideration of space heating and cooling only, instead of the whole heat, power and climatisation requirements of a household, the neglection of mechanical transformation energy (which would be necessary if pressure drops were considered) and the assumption that all heat from a supply system is supplied by one specific process. As a general rule these simplifications should allow a basic but thermodynamically correct assessment of the considered supply systems.

4.2 Results of comparative transformability assessment for examples from heating and cooling

4.2.1 Heating systems

In the following section, the results of an assessment of seven different supply systems for heating and cooling are discussed. The underlying calculations and the energy system models that are the basis for this comparison can be found in appendices A10 on page 143 ff. and A11 on page 152 ff.

Various assessment parameters could be discussed for the purpose of comparison with the transformability ratio and transformation energy efficiency. To keep the evaluation strongly focused, only some of the most promising thermodynamic ratios have been compared. Thus, the very common evaluation using specific CO_2 emissions will not be performed as it is strongly influenced by the choice of fuel, e.g. the specific CO_2 emissions are $202\,\text{g/kWh}_\text{F}$ for natural gas and $404\,\text{g/kWh}_\text{F}$ for lignite (Machat and Werner, 2007). The specific CO_2 emissions associated with the combustion of biomass strongly depend on the way how the biomass is grown, harvested and stored. Since all chemical fuels are associated with a transformability of $100\,\%$, the transformability analysis is no competition but an addition to the greenhouse gas emission evaluation.

Another common parameter for the assessment of energy supply systems which will not be considered is the primary energy factor that is a measure for the amount of fossil energy used to provide one unit of the considered fuel. It is influenced by mining, transport and the type of the fuel considered. It is especially inconvenient for a thorough thermodynamic analysis that the primary energy factor of renewable energy is set to a value near zero. This implies that renewable energies can be used inefficiently without having a negative effect, in spite of the fact that the growth of biomass at least requires the use of fertile land and is therefore limited. Additionally, the evaluation of heat from combined heat and power is based on a power bonus system, which can decrease the primary energy factor associated with heat below a value of zero. The neglection of renewable primary energy and the problematic evaluation of heat from combined heat and power lead to the conclusion that the primary energy factor cannot be considered fully satisfactory to allow a grounded, universal and quantitatively correct assessment of energy systems.

As a consequence, the exemplary thermodynamic evaluation will be limited to four ratios, which are summarized in Table 4.1. To obtain an impression of the applied boundary definition, see

4 Application of the transformability evaluation and analysis method

figures in appendix A10 on page 143. The assessment ratios displayed are the transformation energy efficiency η_τ, transformability ratio ξ, exergetic efficiency ε and the average fuel related coefficient of performance[1] $COP_{a,cF}$. In addition to the actual transformation energy efficiency and transformability ratio, the maximum transformability ratio[2] ξ^{mx} is given in order to allow a more comprehensive discussion of the results. The process design evaluation using the maximum transformability ratio assumes a transformation energy efficiency of those designs of $100\,\%$.

The common supply target requires a heat input of $12\,kW$ at $295\,K$ at a reference temperature of $275\,K$.

Table 4.1: Results of the evaluation of exemplary heating systems

Data	η_τ	ξ	ξ^{mx}	ε	$COP_{a,cF}$
Natural gas condensing boiler	94%	7%	7%	6%	0,94
Geothermal heat source supply system with a forward flow temperature of 80 °C and a return flow temperature of 50 °C	84%	28%	36%	24%	12,30
Electrical compression heat pump with an evaporation temperature of 10°C (heatsource: ground) and a condensation temperature of 36,85 °C (310 K)	71%	13%	60%	9%	1,42
District heating from a block heat and power plant with a forward flow temperature of 85 °C and a return flow temperature of 42 °C	81%	16%	31%	13%	2,09

The $COP_{a,cF}$ is the ratio which is easiest to calculate but also the least accurate, as it neglects all nonchemical exergy inputs. As a result, the $COP_{a,cF}$ of the geothermal heat source is more than twelve times higher than that of the condensing boiler, which could be wrongly interpreted as a higher thermodynamic improvement than it really is. The problem of the $COP_{a,cF}$ is thus its imperfection in regard to a correct quantification of the benefits of one technology over another if not only flows

---

[1] The fuel related coefficient of performance $COP_{a,cF}$ is the ratio of the considered average required energy output to the average input of combustible fuel into the energy supply system within a year.
[2] The maximum transformability ratio can be defined as the transformability ratio of a process operating at the considered parameters but with a transformation energy efficiency of 100 % and without avoidable pressure losses. The maximum transformability ratio is therefore an indicator how well the process type is suited to provide the required energy demand.

of combustible fuels but all thermodynamically relevant flows are of interest. The advantage is its simple calculation and that it provides a good estimate of the resulting ranking according to exergetic efficiency and transformability ratio. Additionally, it cannot be interpreted on its own as an indicator of improvement potential since its values can exceed 100%.

Exergetic efficiency allows a quantitatively correct evaluation and ranking of all considered technologies but remains intransparent to the causes of a given value since a value of exergetic efficiency is influenced by quantitative and qualitative effects alike. *The improved transparency is the major benefit of the transformability assessment.* The advantage of the comparative transformability evaluation becomes apparent when comparing the natural condensing boiler with the electrical compression heat pump. While the exergetic efficiency of the heat pump system is only about $1,4$ times higher than that of the boiler system, its transformability ratio is more than two times higher. This implies that the heat pump suffers greater avoidable losses than the boiler, which is quantified in its lower transformation energy efficiency thus giving more space for external improvements[3]. The low transformability ratio of the boiler system is a characteristic of all processes that use only nonthermal transformation energy to provide a thermal product. This fact becomes obvious if considering ξ^{mx} which equals ξ as a result of the exclusive use of high-transformability chemical transformation energy to generate heat by combustion.

The comparison of maximum with real transformability ratio shows that the transformability ratio is not independent of the transformation energy efficiency for all processes but the boiler. This is a side effect of the definition of the transformability ratio as a ratio of the *average* product transformability to the *average* fuel transformability. Since the average transformability of the fuel increases if heat losses are compensated by additional input of combustible fuel while maintaining the input of heat from the environment, nearly all real average fuel transformabilities are higher than the ones for the calculation of ξ^{mx}. The maximum transformability ratio equals the maximum exergetic efficiency that the considered supply system can reach if its design and parameters are not changed. Whether external losses can influence this ratio becomes obvious if comparing the real transformability ratio with the maximum one. If both values are equal, then losses have no influence on fuel composition indicating that the transformability ratio is a measure of the fundamental process suitability. If the value of the maximum transformability ratio is low, such as for the boiler heat supply system, it is an indicator for the need to replace the considered process with a different process using different fuels.

Further comparison of real transformability ratios and maximum transformability ratios shows that none of the considered processes is optimally suited for the considered supply task. However, the difference between maximum transformability ratio and 100% indicates that the processes can theoretically be improved significantly by design modifications (changing the flow chart) or process parameter changes (e.g. changing process temperatures). The most fundamental change that could be performed to increase the maximum transformability ratio is to change all processes in such a way that the average temperature of the provided heat nearly equals the required temperature. Since the transformability of the required heat is only 7%, due to a temperature difference of only 23 K between reference temperature and required supply temperature, even small differences of 20 K

---

[3] A large part of the avoidable losses are caused by a throttling process in the heat pump, which results in an indirect transformation energy loss since additional fuel is required to compensate for the transformability destruction.

4 Application of the transformability evaluation and analysis method

between required temperature and average supply temperature result in maximum transformability ratios below 50 % for all systems but the one using the heat pump. Considering this background, it is e.g. easy to explain why the maximum transformability ratio of the compression heat pump system is higher than that of the geothermal heat supply system. The heat pump supply system has a supply temperature of only 36,85 °C, while the average heat supply temperature of the geothermal heat supply system is 64,78 °C. Due to its lower supply temperature, the ideal heat pump system is better suited for the supply task than the energetically ideal geothermal heat supply system. However, the real transformability ratio of the heat pump supply system is significantly lower than the transformability ratio of the geothermal heat supply system. This is mainly caused by the the high energy losses of the power plant providing electricity for the heat pump, which have to be compensated by additional input of combustible fuel into the heat supply system. The additional fuel input into the heat pump heat supply system shifts the average fuel transformability of the compression heat pump to a value which is larger than the average fuel transformability of the real district heating system with a geothermal heat source.

If comparing district heating based on CHP with district heating using heat from a geothermal source, it becomes apparent that the losses from the CHP process, which through fuel attribution to heat influence the average transformability of heat from CHP, significantly affect the transformability ratio. Still, the district heating system using combined heat and power performs better than the heat pump system for the evaluated examples, although due to its higher supply temperature it has a lower maximum transformability ratio.

4.2.2 Cooling systems

The assessment of cooling systems using the transformability evaluation method is more complex than the evaluation of processes which operate completely above reference temperature since thermal compensation heat flows have to be considered. This becomes obvious in the flow charts presented in appendix A11 on page 152 ff. The way in which thermal compensation heat flows are considered in the calculation of the transformation energy efficiency is discussed in subsection 3.2.3 on page 54 ff. Additionally, all heat flows above reference temperature that are discharged from cooling processes are considered as heat flows at reference temperature, since their transformability is destroyed by the discharge. Certainly, the heat at the real discharge temperatures could be recovered but in this case it would become a useful flow thus requiring consideration in the numerator of the evaluation ratios. As processes that produce heat and cold simultaneously as useful products are very rare, the transformability evaluation of these special processes is not considered here.

Table 4.2 shows the results of evaluation for three fundamentally different cooling supply systems. The target supply system requires a heat extraction of $5\,kW$, which equals a transformation energy input of $5,25\,kW$ at $295\,K$ with a reference temperature of $310\,K$[4].

---

[4]This assumed average daily temperature could be valid for some very hot deserts in summer. It has been chosen in order to obtain a theoretically required transformability of the thermal transformation energy for cooling of approximately 5 %. This required transformability is similar to the required transformability of the thermal transformation energy used for heating of approximately 7 %, while at the same time being realistic for certain areas of the world.

4 Application of the transformability evaluation and analysis method

Table 4.2: Results of the evaluation of exemplary cooling systems

Data	η_τ	ξ	ξ^{mx}	ε	$COP_{a,cF}$
Compression refrigeration machine with heat discharge to air at 325 K and an average cooling temperature of 282,15 K (9°C)	49%	8%	33%	4%	0,81
Direct cooling with 280 K seawater	94%	32%	52%	30%	13,67
Absorption refrigeration machine which uses waste heat at 350 K and provides cooling at T_a 282,15 K (9°C)	62%	19%	24%	12%	9,23

Like for heating systems a ranking according to the $COP_{a,cF}$ would lead to the same result as a ranking based on transformability ratio or exergetic efficiency and again the quantitative relations between the compared systems would be significantly different. An additional drawback of an assessment scale that is not limited from 0 to 100 % is the impossibility to relate the values to a maximum performance. As a conclusion, the $COP_{a,cF}$ appears to be sufficient for ranking the considered processes, while its values are only useful if the evaluation target is to assess effectiveness in regard to combustible fuels.

Obviously, the exergetically best option is the direct seawater cooling. Since all supply systems use chemical transformation energy as well as transformation energy associated with heat flows, the transformability ratio of all systems is dependent on transformation energy losses. The absorption refrigeration machine that is driven by waste heat[5] is significantly better suited to the supply task than the compression refrigeration machine system, which has a transformability ratio that is more than two times lower than the one of the absorption refrigeration machine. However, it is interesting to note that the compression refrigeration machine has a higher maximum transformability ratio than the absorption refrigeration machine system, which implies that essentially the parameters of the compression refrigeration machine are more suitable to fulfill the task. The central reason for this strong deviation of real from ideal value is low electrical efficiency of the average combustion power plant, which is assumed to provide the electricity.

Similarly to heat pumps, refrigeration machines also suffer from comparably low transformation energy efficiencies indicating a larger improvement potential of the process at the given parameters than the direct seawater cooling. The improvement potential is mainly found in the area of power generation

---

[5]Waste heat can be defined as heat the use of which has no influence on the operation of the process generating it. The parameters and transfers of the waste heat generator as well as its operation are therefore not influenced by the use of the waste heat.

and component replacement of the throttles by expanders.

Since the major advantage of the separate evaluation by transformability ratio and transformation energy efficiency is the greater transparency in comparison with the exergetic efficiency, all supply systems have been considered at three different reference states, so that the effect of a changing reference state on the considered properties becomes visible.

4.3 Influence of reference temperature on the evaluation of thermal supply systems

The definition of the reference state is the basic assumption for the calculation of exergy values. Therefore it is important to assess the influence of varying reference state conditions on the results of steady-state exergetic evaluation and analysis. Rosen and Dincer (2004) have performed a sensitivity analysis of thermal and physical exergy to changes in reference state. They defined the exergetic sensitivity σ as a function of the exergy flows $\dot{E}$ of the considered flow at average reference temperature T_0 and at the exact reference temperature $T_0 + \Delta T_0$:

$$\sigma = \frac{\dot{E}\left(T_0 + \Delta T_0\right) - \dot{E}\left(T_0\right)}{\dot{E}\left(T_0\right)}$$

For steady-state conductive heat flows they obtained the following expression:

$$\sigma^Q = \frac{\Delta T_0}{T_0 - T}$$

Since the denominator of the ratio is usually small for heat flows required for domestic heating and cooling, fluctuations of the reference temperature can in principle have a large impact on the required exergy values. However, this expression also implies that the use of average reference temperatures for the period in which the considered system operates, provides exact values for the average exergy associated with a considered flow, which is demonstrated for an example in appendix A12 on page 162[6].

The influence of changes in reference temperature on evaluation results obtained with transformability evaluation shall shortly be discussed based on the assessment of examples discussed in section 4.2 for two more reference temperatures.

Chemical exergy is considered to be equal to the higher heating value. Therefore it is considered independent of the reference temperature and composition. Since of the other types of transformation energy only thermal transformation energy is considered in the comparative assessment, the only reference parameter that can influence the given results is the reference temperature. Tables 4.3 and 4.4 summarize the assessment results for the considered supply systems using three different reference temperatures for the evaluation of heating systems and another three reference temperatures for cooling.

---

[6]The equivalence of the average exergy values associated with a heat flow and the exergy associated with the heat flow at average temperature is only valid for the assumption of steady-state heat flows and only for the use of the average temperature of the period in which the system is operating. Since in reality the magnitude of heat losses depends on the outside temperature further investigations can be based on a dynamic exergy analysis.

4 Application of the transformability evaluation and analysis method

These tables show that a comparison of heating and cooling technologies at different reference temperatures is problematic. The strong influence of reference temperature on transformability ratio suggests that the results of such a comparison would be significantly less meaningful, since not only process parameters and energy losses influence the actual degree of suitability but also the reference environment at which the different processes are evaluated.

Considering Table 4.3 it becomes apparent that for most technologies the fuel-related coefficient of performance is independent of reference temperature. Only for the combined heat and power supply system the $COP_{a,cF}$ increases with increasing temperature, which is sensible since the fuel attribution to heat is a function of the exergy associated with the heat flow and consequently also of reference temperature. More fuel is therefore attributed to a heat flow for lower reference temperatures resulting in lower fuel related coefficients of performance.

4 Application of the transformability evaluation and analysis method

Table 4.3: Results of the evaluation of exemplary heating systems at different reference states

Supply system	T_0	Shortname	η_τ	ξ	ε	$COP_{a,cF}$
Heat supply system based on a condensing boiler	265 K	NG-cond. Boiler (T0l)	93,86%	10,17%	9,54%	0,94
	275 K	NG-cond. boiler	93,86%	6,78%	6,36%	0,94
	285 K	NG-cond. Boiler (T0h)	93,86%	3,39%	3,18%	0,94
Heat supply system based on geothermal district heating	265 K	Geothermal (T0l)	83,86%	37,77%	31,67%	12,30
	275 K	Geothermal	83,86%	28,05%	23,52%	12,30
	285 K	Geothermal (T0h)	83,86%	15,83%	13,28%	12,30
Heat supply system based on a ground-source heat pump	265 K	EL-HP: Ground (T0l)	70,65%	19,20%	13,56%	1,42
	275 K	EL-HP: Ground	70,65%	13,24%	9,36%	1,42
	285 K	EL-HP: Ground (T0h)	70,65%	6,86%	4,85%	1,42
Heat supply system based on district heating from a block combined heat and power plant	265 K	Conv. DH (T0l)	81,47%	21,12%	17,20%	1,85
	275 K	Conv. DH	81,36%	15,66%	12,74%	2,09
	285 K	Conv. DH (T0h)	81,24%	8,89%	7,22%	2,41

4 Application of the transformability evaluation and analysis method

The transformation energy efficiency, which equals energy efficiency for the considered supply systems, is independent of reference temperature, as is the required transformation energy. Only for the combined heat and power supply system a decrease of the transformation energy efficiency with increasing reference temperature is notable, which is an effect of the dependence of the fuel attribution factor of the reference temperature. However, the very weak influence on the evaluation result allows to consider the transformation energy efficiency of the heat supply from CHP as quasi constant. As a consequence it becomes obvious that the transformation energy efficiency definition in connection with the exergy-based fuel attribution as discussed in section 3.4 on page 62 ff. results in an exergy-based quantitative evaluation ratio that is by itself nearly independent of small variation in reference temperature and thus a good indicator of a degree of external sophistication that is only dependent on the process, in contrast to the $COP_{a,cF}$, which is also significantly dependent on reference temperature for heat from combined heat and power.

It is apparent that the transformability ratio is strongly dependent on reference temperature. Since exergetic efficiency can be considered a product of transformability ratio and transformation energy efficiency, the separate evaluation of these two properties shows that only the qualitative ratio is influenced by the reference temperature while the quantitative ratio remains relatively independent. This is illustrated by Figure 4.1, which represents a more transparent way of a graphical exergy-based assessment, than would be possible with the exergy concept alone. As a consequence, the transformability assessment method allows to identify the considered supply system at a different reference state by its transformation energy efficiency while a purely exergetic evaluation cannot provide this identification potential.

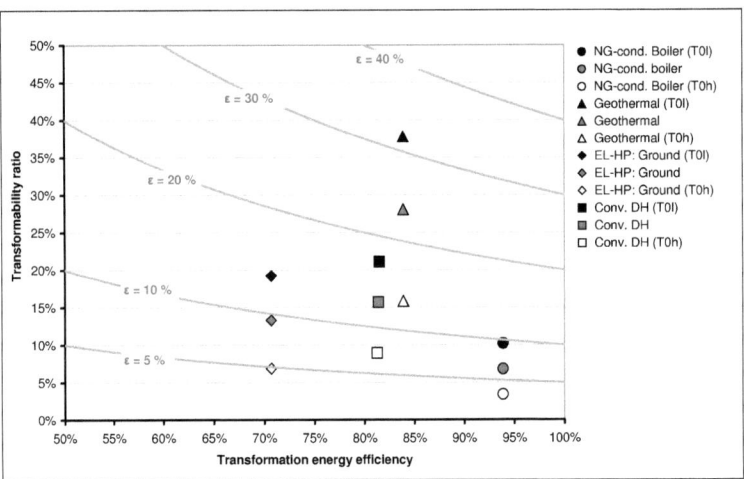

Figure 4.1: Transformation energy efficiency - Transformability ratio diagram for heat supply systems at different reference temperatures

Table 4.4: Results of the evaluation of exemplary cooling systems at different reference states

Supply system	T_0	Shortname	η_τ	ξ	ε	$COP_{a,cF}$
Compression refrigeration machine system for domestic cooling	300 K	Comp. refrigeration machine: air (T0l)	48,43%	2,83%	1,37%	0,81
	310 K	Comp. refrigeration machine: air	49,25%	8,36%	4,12%	0,81
	320 K	Comp. refrigeration machine: air (T0h)	50,04%	13,72%	6,86%	0,81
Direct cooling system using seawater	300 K	DC seawater (T0l)	93,56%	13,28%	12,42%	13,67
	310 K	DC seawater	93,76%	31,55%	29,58%	13,67
	320 K	DC seawater (T0h)	93,94%	43,51%	40,88%	13,67
Absorption refrigeration machine cooling system using waste heat	300 K	Ab.- refrigeration machine: air (T0l)	62,03%	5,52%	3,43%	9,23
	310 K	Ab.- refrigeration machine: air	62,03%	18,90%	11,73%	9,23
	320 K	Ab.- refrigeration machine: air (T0h)	62,03%	36,68%	22,75%	9,23

4 Application of the transformability evaluation and analysis method

Table 4.4 on the previous page shows the results of the evaluation of cooling technologies at different reference states. Although additionally to the transformation energies associated with the heat flows the effective compensation heat flow has to be considered, the transformation energy efficiency remains an indicator only weakly influenced by reference temperature while the transformability ratio is a very strong function of reference temperature. Here, transformation energy is a unique property different from energy, as the source of work that can be obtained from thermal interaction of environment and flow under consideration is the environment. Thus, the transformation energy that is associated with heat extracted from the supply target at a temperature below reference temperature changes with changing reference temperature. This dependence of the transformation energy of the product causes a slight dependence of transformation energy efficiency on reference temperature.

In general the transformability ratio decreases with decreasing reference temperature, as the reference temperature approaches the temperature of the supply target making the considered application more and more obsolete. Interestingly, the transformability ratio of the absorption cooling system increases significantly stronger than the transformability ratio of the direct cooling system with seawater. The observed effect is a result of the decreasing transformability of the driving heat of the absorption refrigeration machine, while the transformability associated with the cool seawater is increasing with increasing reference temperature.

Figure 4.2 shows the results in a graphical way. It becomes apparent that the transformation energy efficiency is a weak function of reference temperature for two of the three cooling systems considered.

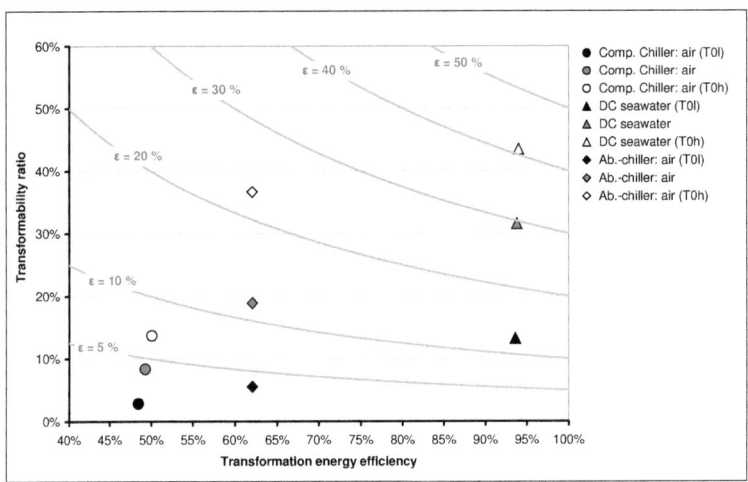

Figure 4.2: Transformation energy efficiency - Transformability ratio diagram for cooling systems at different reference temperatures

The different dependence of the transformation energy efficiency on reference temperature for the considered examples is an effect of the inclusion of the effective compensation heat flows into

4 Application of the transformability evaluation and analysis method

the transformation energy efficiency equations. While the required transformation energy has the same dependence on reference temperature for all evaluated processes, the influence of the effective compensation heat flows depends on the summands included in its calculation and whether it is considered in the numerator or the denominator of the efficiency equation.

For the compression refrigeration machine, the effective compensation heat flow is a sum of the compensation heat flow associated with the heat extraction from the supply target and the ideally required heat discharge from the refrigeration machine and has to be considered in the *denominator* of the transformation energy efficiency definition. In contrast to that, the transformation energy efficiency of the absorption refrigeration machine includes an effective compensation heat flow in the *numerator* that is also a sum of the compensation heat flow associated with the heat flow from the supply target and an ideally required heat discharge. Interestingly, this leads to a transformation energy efficiency which is totally independent of reference temperature. However, the inclusion of the effective compensation heat flow in the numerator does not guarantee an independence of the transformation energy efficiency on reference temperature. This becomes obvious when considering the effective compensation heat flow of the direct seawater cooling system which also has to be included in the numerator but is a sum of two compensation heat flows and does not include an ideally required heat flow. In this case, the reference temperature still has an influence on transformation energy efficiency.

These examples show that the exact values and definitions of the effective compensation heat flows define the influence of reference temperature on transformation energy efficiency . However, all transformation energy efficiencies of the considered processes are at best weak functions of the reference state and can be considered constant for small deviations of reference state from a given starting value. The calculation of transformation energy efficiency for the discussed examples can be found in appendix A11 on page 152 ff.

4.4 Effects of heat losses and pressure drops on the evaluation of some basic processes

Transformability analysis of basic processes and single components is, like the comparative transformability evaluation discussed in the previous section, based on transformation energy efficiency and transformability ratio. However, some significant differences exist between the application of the transformability concept to comparative assessment and to thermodynamic analysis. The comparative transformability assessment is based on a simplified energy system approach with a supply target and a balance boundary into which only primary energy enters, furthermore mechanical exergy is neglected. Process analysis on the other hand requires a more comprehensive and less simplified approach to evaluation, since it has to include mechanical transformation energy and therefore also requires significantly more data and steps of calculation. Additionally, it also has a different focus. Instead of quantifying the differences between alternative systems that can provide a desired product (heating or cooling), analysis aims at the identification of improvement potential within a considered balance boundary.

To be able to evaluate the transformability analysis, a set of some common thermodynamic processes has been modelled. The resulting equations for transformation energy efficiency, average in- and output transformabilities, transformability ratios and exergetic efficiencies have been collected in Tables A.16, A.17, A.18 and A.19 on page 166 ff. Due to the amount of data that can be easily calculated for these processes using a software implementation of the equation of states, e.g. from NIST (2007), no numerical representation of the assumptions will be given as it would significantly increase the volume of this dissertation. A set of 25 calculations has been performed for some examples of the considered processes. All considered input mass flows are assumed to be flows of dry air at reference pressure. The calculation was performed in four steps. First, the ideal values were calculated, then the effects of a pressure drop. After assessing the mechanical transformation energy decrease that is a part of the fuel term in the transformation energy efficiency, the influence of a heat loss was considered. Finally, the simultaneous influence of the chosen heat loss and the chosen pressure drop in transformation energy efficiency and transformability ratio have been assessed.

To obtain an impression of the impact of heat losses and pressure losses on the evaluation ratios, it has been assumed that the heat losses equal the total decrease in mechanical transformation energy within the process. Additionally, it has been checked that all processes have attained a transformation energy efficiency of 100% for the case of no losses and pressure drops. For processes which even in the energetically ideal case are not characterized by a transformability ratio of 100%, the ideal process values have been given separately. All other processes have ideal transformation energy efficiencies of 100% and maximum transformability ratios of 100%. Although the basic data for temperatures, pressures, resulting work flows, enthalpies, entropies as well as exergies, transformation energies and compensation heat flows are not given to limit the volume of this work, the numerical results of the calculation are provided to allow the discussion of the effects of pressure and heat losses in greater detail.

The general analysis was performed to answer the question:

How do pressure drops and conductive heat losses influence transformation energy efficiency and transformability ratio?

Due to the exemplary nature of this first numerical application of the transformability analysis for the analysis of processes, all results are only indicators. A more fundamental discussion of the effects can be based on the equations provided in Tables A.16 on page 166, A.17 on page 167, A.18 on page 168 and A.19 on page 169 if necessary. Table 4.5 summarizes the results of a series of calculations using the model equations provided in the appendix A13 on page 164 ff..

Table 4.5: Results of the evaluation of basic processes

Process	Influence	η_τ	ξ	ε
Heat exchanger	Pressure losses	100,00%	81,91%	81,91%
	Pinch	100,00%	93,56%	93,56%
	Pressure losses and pinch	100,00%	76,64%	76,64%
Boiler	Ideal process	100,00%	19,04%	19,04%
	Pressure losses	100,00%	13,68%	13,68%
	Heat losses	70,12%	19,04%	13,35%
	Pressure losses and heat losses	76,54%	13,68%	10,47%
Compression heat pump	Pressure losses	100,00%	56,35%	56,35%
	Heat losses	87,19%	64,63%	56,36%
	Pressure losses and heat losses	88,65%	44,25%	39,23%
Compression refrigeration machine ($T_l < T_0 < T_h$)	Ideal process	100,00%	71,87%	71,87%
	Pressure losses	100,00%	36,02%	36,02%
	Heat losses	89,49%	40,24%	36,01%
	Pressure losses and heat losses	89,49%	26,85%	24,03%
Compression refrigeration machine 7 ($T_l < T_h < T_0$)	Pressure losses	100,00%	68,87%	68,87%
	Heat losses	91,61%	75,11%	68,81%
	Pressure losses and heat losses	92,26%	56,90%	52,49%
Heat engine	Ideal process	100,00%	73,78%	73,78%
	Pressure losses	100,00%	48,77%	48,77%
	Heat losses	68,17%	33,00%	22,50%
	Pressure losses and heat losses	75,86%	19,60%	14,87%
Expander (inflow at T_0)	Adiabatic - heat loss	82,34%	80,50%	66,29%
	Isothermal - heat loss	66,29%	100,00%	66,29%
Compressor (inflow at T_0)	Adiabatic - heat loss	86,21%	88,90%	76,64%
	Isothermal - heat loss	74,79%	100,00%	74,79%

4.4.1 Heat Exchanger

The effects of the considered changes influence only the transformability ratio, while the transformation energy efficiency remains constant at $100\,\%$. This shows that with the given equations, the transformability analysis shows a large amount of transformability destruction in the heat exchange process, either by pressure loss or due to a temperature difference between the feed and the product flow. This result is consistent with the fact that instead of heat loss, a pinch temperature difference has been assumed.

4.4.2 Boiler

The ideal boiler cannot reach a transformability ratio equal to $100\,\%$, since the transformation of chemical transformation energy into thermal transformation energy by combustion is always accompanied by transformability destruction. Pressure losses only influence transformability ratio, while heat losses affect only the transformation energy efficiency. Such a clear separation of effects is caused by simple nature of the process, the fuel input of which is associated with a constant average transformability. As a consequence, if heat and pressure losses occur together, the transformability ratio is equal to that with the pressure losses only. The transformation energy efficiency on the other hand changes if in addition to a heat loss also a pressure loss occurs. This can be explained by the larger amount of transformation energy considered in the fuel term of the efficiency equation. Thus, a given heat loss has a lower impact on transformation energy efficiency if also pressure losses are present, which increase the fuel term while at the same time the effective compensation heat increases the product term.

The exergetic efficiency of the boiler with pressure and with heat losses alone is very similar, thus showing that exergetic efficiency cannot provide information that the transformability ratio and transformation energy efficiency can. If heat losses occur, the low transformation energy efficiency shows external improvement potential while in the case of pressure losses the decreased transformability ratio indicates internal

improvement potential. Generally, the low transformability ratio of the boiler indicates its poor suitability for the considered task of providing low-temperature heat by using combustible fuels.

4.4.3 Compression heat pump

The transformability ratio of the compression heat pump shows a common characteristic with the transformability ratios of the boiler and the heat exchanger - pressure losses only influence the transformability ratio. In contrast to the boiler, heat losses influence both the transformability ratio and transformation energy efficiency. This can easily be explained if considering that heat losses have to be compensated by additional input of power, while the product heat flow remains the same. As a consequence, the average transformability of the fuel increases resulting in a decrease of transformability ratio. Like in the case of the boiler, pressure losses and heat losses alone result in nearly identical values for exergetic efficiency. This confirms the interpretation that heat losses have to

4 Application of the transformability evaluation and analysis method

be compensated by additional fuel input. Otherwise, due to the lower value of thermal transformation energy in comparison to mechanical transformation energy, a different exergetic efficiency would have been expected. However, the presence of pressure losses alone does not decrease the transformation energy efficiency, thus indicating that the improvement potential of the considered process is mainly internal.

4.4.4 Compression refrigeration machines

The compression refrigeration machine that operates between temperatures above and below reference temperature has a transformability ratio lower than $100\,\%$ even if operating ideally. This has its origin in the assumption that the exiting heat flow is discharged directly to the environment and is thus associated with a transformability of $0\,\%$ instead of the transformability matching its exit temperature. Since the ideally required heat flow, which affects the effective compensation heat, is a function of the low and high temperature of the process, a part of the average fuel transformability is destroyed even in the energetically ideal case.

Pressure losses, like for the previously considered processes, influence only transformability ratio, while heat losses like for the heat pump lead to a transformation energy loss, resulting in a decrease of the transformation energy efficiency. An interesting result is the fact that heat and pressure loss combined yield a transformation energy efficiency that is identical with the one of the pressure loss only. So in contrast to the heat pump, no influence of the pressure losses on transformation energy efficiency can be noted. Considering Table A.18 on page 168, this can be explained with the positive sign of the effective compensation heat, which results in a subtraction of the mechanical compensation heat flow difference from the mechanical transformation energy input, thus nearly nullifying its influence.

The results of the evaluation of the refrigeration machine operating completely below reference temperature are very similar to the results obtained for the heat pump if heat and pressure losses are considered or if energetically ideal operation is considered. While at energetically ideal operation all presented ratios are $100\,\%$, pressure losses decrease the transformability ratio, while heat losses decrease transformability ratio and transformation energy efficiency. A combination of heat and pressure loss results in a change of both evaluation properties.

The impact of heat losses on exergetic efficiency of compression refrigeration machines is equal to the one of pressure losses, thus indicating that heat losses lead to an increased fuel consumption of the process.

4.4.5 Heat engine

The heat engine is significantly different from the previously considered processes, as it is a process with a nonthermal useful energy flow. For a heat engine with a condensation temperature above reference temperature, the maximum transformability ratio must be lower than $100\,\%$, since even in the ideal process transformability is destroyed by discharging heat a temperature above reference temperature to the reference environment. Like in all of the previously discussed processes, a pressure loss only decreases the transformability ratio. Heat losses decrease transformation energy efficiency

4 Application of the transformability evaluation and analysis method

and transformability ratio. This can be explained if considering the average output transformability equation in Table A.17 on page 167. Since with heat losses the generated work flow decreases while the effective compensation heat flow remains constant, the average product transformability is decreased by heat losses. Consequently, heat losses have a stronger influence on the exergetic efficiency than pressure losses, since they decrease the amount and the value of the generated product, while pressure losses only result in an additional fuel transformation energy term.

4.4.6 Expander

In contrast to the previously considered processes, the expander uses transformation energy associated with mass flows directly. This means that the mechanical transformation energy decrease of the mass flow is required to operate the process, while for all previous processes mechanical transformation energy decreases, which had to be considered in the fuel term, had always been unwanted side effects. Thus, it appears to be sensible to consider adiabatic and isothermal operation of the expander if heat losses occur. Heat losses in an expansion process starting at reference temperature can only occur after gas expansion in the generator, since in principle the gas is cooling below reference temperature in the adiabatic case. These heat losses directly decrease the generated work flow, while not influencing gas properties. If considering the isothermal process, according to the equation in Table A.17 the effective compensation heat flow has a value of zero. Thus, no thermal transformation energy is exiting or entering an isothermal expansion process at reference temperature, allowing the average input transformability and the average output transformability to remain unchanged by a decrease in power production. However, the heat losses in an adiabatic process result in a decrease in transformation energy efficiency and transformability ratio, since with the decrease of generated power the share of thermal transformation energy in the product increases, thus decreasing the average product transformability. It is interesting to note that the exergetic efficiency for a given heat loss is independent of the type of the expansion process considered. This can be explained by the same decrease in power production for a given heat loss, independent whether the process considered is adiabatic or isothermal.

4.4.7 Compressor

In principle the compressor is affected by heat losses in a similar way to the expander. For adiabatic compression heat losses influence transformation energy efficiency and transformability ratio, while for isothermal processes only transformation energy efficiency is affected. The major difference is the fact that exergetic efficiency for a given heat loss depends on the type of process. The isothermal process requires in the ideal case less power to achieve a given change in pressure than the adiabatic process. Therefore, a given heat loss and the resulting increase in work flow input, which is added to the ideally required work flow input, increases the total power input by a higher percentage for the isothermal process than for the adiabatic one.

4 Application of the transformability evaluation and analysis method

4.4.8 Summary

It has been shown that although the considered pressure losses can result in negative values of the specific transformation energy associated with the outgoing flow, this does not lead to inconsistencies. On the contrary, calculations have shown that the mechanical transformation energy influence is essentially independent of absolute pressure and only depends on the relative amount of pressure loss. This is essentially caused by the way the consideration of mechanical transformation energy increases or decreases in the efficiency definitions instead of taking inputs and outputs of mechanical transformation energy into account separately.

All processes that use mass flows indirectly and require only thermal transformation energy for operation display the same behavior concerning pressure losses: a decrease in transformability ratio only. This indicates that the pressure losses have no effect on the external performance of the considered processes. However, decreases in mechanical transformation energy can also lead to an increased fuel consumption and thus to the necessity to discharge heat flows in order to maintain target process parameters. In such cases, pressure losses would indirectly influence transformation energy efficiency.

The examples show that whenever transformation energy efficiency is nonideal, it is a clear indicator that heat losses in the system are present. This supports the interpretation of transformation energy efficiency as an indicator of the degree of external sophistication as has been discussed in subsection 3.2.4 on page 57 ff. The evaluation of boilers using transformability analysis increases transparency significantly, since transformability ratio is independent of heat losses. In most of the other processes a clear identification of the origin of a specific nonideal transformability ratio value is not possible using these ratios alone. Nonetheless, the transformability analysis makes it possible to identify and to quantify the central problem of the considered system, since the two ratios allow to assess whether external inefficiencies or suboptimal process suitability have the strongest influence on the overall exergetic performance. Consequently, it appears that transformability analysis could be an interesting and useful extension to conventional exergy analysis, providing first indications to the improvement potentials of the considered system and a new perspective on process performance.

4.5 Example - Analysis of a vapor-compression cascade refrigeration machine

In this section the results of a transformability analysis of a vapor-compression cascade refrigeration machine are discussed. The calculations that lead to the results are presented in appendix A14 on page 165 ff.

Figure 4.3 shows the flow chart of the considered cycle.

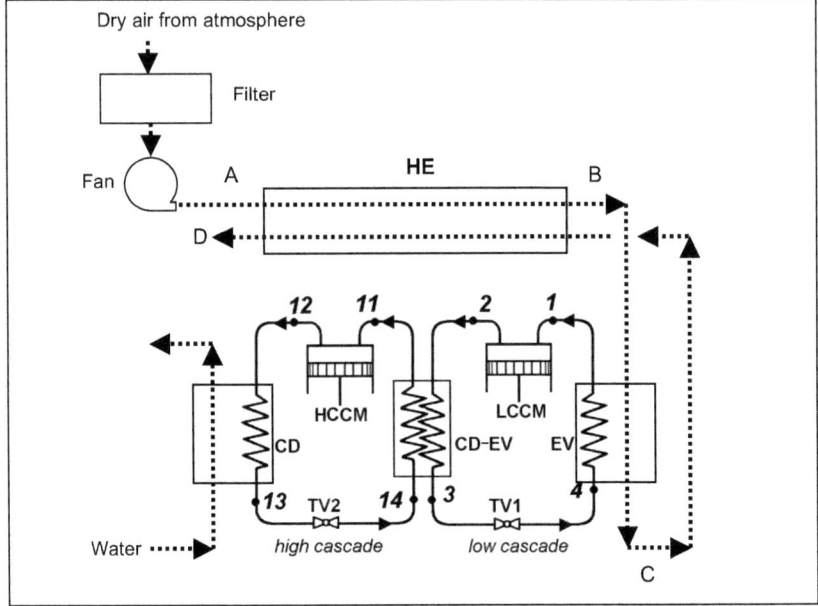

Figure 4.3: Vapor-compression cascade refrigeration machine

The purpose of the vapor-compression cascade refrigeration machine is to cool down air from state B to state C. Heat from the high cascade condenser [CD] is finally discharged to the environment. Air at state D is not used further. The results of the analysis are presented in table 4.6.

4 Application of the transformability evaluation and analysis method

Table 4.6: Results of the transformability analysis of a vapor-compression cascade refrigeration machine

Process	η_T	ξ	ε
Low cascade - evaporator [EV] $(4 \to 1/B \to C)$	100,00%	30,24%	30,24%
Low cascade - compressor [LCCM] $(1 \to 2)$	90,59%	49,30%	44,66%
Intercascade condenser/evaporator [CD-EV] $(2 \to 3/14 \to 11)$	100,00%	59,81%	59,81%
High cascade - compressor [HCCM] $(11 \to 12)$	96,35%	57,19%	55,11%
Low cascade - throttle [TV1] $(3 \to 4)$	100,00%	84,94%	84,94%
High cascade - throttle [TV2] $(13 \to 14)$	100,00%	78,92%	78,92%
Total process	51,16%	28,04%	14,35%

In contrast to exergy, the definitions of thermal transformation energy and transformability are different for temperatures above and temperatures below reference temperature. Thus, it was important to exemplary analyze a process that operates below as well as above reference temperature using transformability analysis. The consistent evaluation of the considered cascade refrigeration process indicates that transformability analysis is as universally applicable as exergy analysis. Using the understanding gained from the general analysis of processes performed in the previous section, the results can easily be interpreted.

Since the transformation energy efficiencies of the low-cascade evaporator [EV], the intercascade condenser/evaporator [CD-EV] and the throttles are 100%, they are considered energetically lossless. The low transformability efficiency of the low-cascade evaporator indicates that the parameters of this process have the greatest potential for improvements. The low-cascade compressor [LCCM] has a significantly lower transformability ratio in comparison to the high-cascade compressor [HCCM] as a consequence of an isentropic efficiency of 63% instead of 73%. All considered process components have a high transformation energy efficiency, indicating a high external sophistication of the parts used. However, the transformation energy efficiency of the total process is significantly lower than the transformation energy efficiencies of the analyzed components. It can therefore be concluded that the energetically problematic components are not those considered. The cooler [CD] is a dissipative component, which means that no product can be defined which would be in accordance with the rules for product definition laid out in subsection 3.2.2 on page 51 ff (Tsatsaronis et al.,

4 Application of the transformability evaluation and analysis method

2007). Thus, the low transformation energy efficiency can be explained with the presence of the cooler in the total process. While energy efficiency cannot be used to sensibly assess the thermal performance of refrigeration machines, exergetic efficiency provides a means for its evaluation. Using transformation energy efficiency, it can be pointed out additionally that the considered components cannot be improved by means external to these components, such as the use or mitigation of waste heat. The comparably low transformability ratio of the components indicates that process design improvements are the most promising way for process optimization.

In the following, the results of the transformability analysis are evaluated considering the transformability values given in Table A.27, which have been calculated based on data and equations presented in appendix A14 on page 165 ff.

Table 4.7: Effective thermal transformabilities, average transformabilities and effective compensation heat flows in the vapor-compression cascade refrigeration machine

Process	$\dot{Q}_i^{id}$	$\Delta \dot{Q}_i^*$	τ_F^H	τ_P^H	$\tau_{a,F}$	$\tau_{a,P}$
		$\frac{kJ}{s}$				
Low cascade - evaporator [EV] ($4 \to 1/B \to C$)	-	-7,78	31,58%	25,23%	52,80%	15,97%
Low cascade - compressor [LCCM]($1 \to 2$)	-1,86	-6,53	-	20,53%	75,49%	37,22%
Intercascade condenser/evaporator [CD-EV] ($2 \to 3/14 \to 11$)	-	-4,31	18,62%	14,54%	18,62%	11,14%
High cascade - compressor [HCCM] ($11 \to 12$)	-1,62	13,58	0,86%	-	40,04%	22,90%
Low cascade - throttle [TV1] ($3 \to 4$)	-	0,16	-	81,58%	96%	81,58%
High cascade - throttle [TV2] ($13 \to 14$)	-	18,24	-	16,58%	21,01%	16,58%
Total process	-14,20	2,63	-	25,23%	89,96%	25,23%

The effective thermal transformability of the fuel τ_F^H in the low-temperature heat exchanger is only 6 % higher than the effective thermal transformability of the product τ_P^H, thus the difference in average temperature of fuel and product is not the problem of the component that leads to the low transformability ratio. If considering the final average transformabilities of fuel $\tau_{a,F}$ and product $\tau_{a,P}$, it becomes obvious that while the average fuel transformability is more than 50 % larger than the effective thermal transformability, the average product transformability is more than 50 % lower

4 Application of the transformability evaluation and analysis method

than the effective thermal transformability of the product. Thus, internal effects, which in the case of the considered component can only be pressure losses, have to account for the low transformability ratio of the low-cascade evaporator. I.e., by considering the average transformability values of fuel and product transformation energy flows, the optimization potential has been identified and can be communicated as a problem of using high-value mechanical transformation energy in a component providing only a thermal product.

Considering Table 4.7 further, it becomes apparent that an effective thermal transformation energy flow is a part of the product for the low-cascade compressor, while an effective thermal transformation energy flow with a very low effective thermal transformability is a part of the fuel for the high-cascade compressor. While the effective thermal transformation energy decreases the average product transformability in the low-cascade compressor, it decreases average fuel transformability in the high-cascade compressor. In oder to improve the efficiency of the low-cascade compressor, the comparably high effective thermal transformability should be decreased. This could be achieved by modifying the compressor in such a way that the average temperature of the mass flow between inlet and exit approaches reference temperature. As long as transformation energy efficiency remains constant or improves with such a modification, the exergetic efficiency would also increase.

This discussion shows that the transformability analysis can help to identify weaknesses clearer than using exergetic efficiency alone at the expense of an increased complexity of calculation. Considering the high transformability ratios of the six components, it becomes obvious that simple external optimization measures such as insulation or leak sealing have no significant potential to improve the considered process. The major problem of all components is a suboptimal suitability, which in conjunction with the high transformation energy efficiency indicates a need for improvement of in- and outflow parameters, e.g by decreasing temperature pinches and pressure drops in the heat exchangers or by changing output temperatures of the compressors. In contrast to the considered components, the low transformation energy efficiency of the total process requires significant external optimization. Since no indication of such an improvement potential is given in the transformation energy efficiency of the components, the improvement potential must lie in optimization of the cooler [CD] which cannot be sensibly evaluated on its own. One way to improve the transformation energy efficiency of the process is the use of the heat discharged from the cooler in another process. The transformability analysis therefore allows a more differentiated view on specific improvement potentials of the considered process than the use of exergetic efficiency. The application to further processes can show with time whether the more differentiated results of transformability analysis justify the greater computational effort required.

4.6 The ExergyFingerprint - a transformability-based graphical assessment tool

Fraunhofer Institute for Environmental, Safety, and Energy Technology UMSICHT has developed a tool for the assessment of energy supply scenarios based on the transformability and transformation energy method, which has been named ExergyFingerprint (Jentsch et al., 2009). The ExergyFingerprint allows a graphical characterization of demand and supply scenarios in a two dimensional way, making every ExergyFingerprint characteristic for a given (transformation) energy supply and demand

4 Application of the transformability evaluation and analysis method

scenario[8]. Figure 4.4 shows the ExergyFingerprint for an average german household, to which heat is supplied by the condensing boiler heat supply system discussed in section 4.2. The data, based on which the demand side has been characterized, can be found in Table A.30 on page 173, while data for the supply side can be found in Tables A.5 on page 144 and A.6 on page 145.

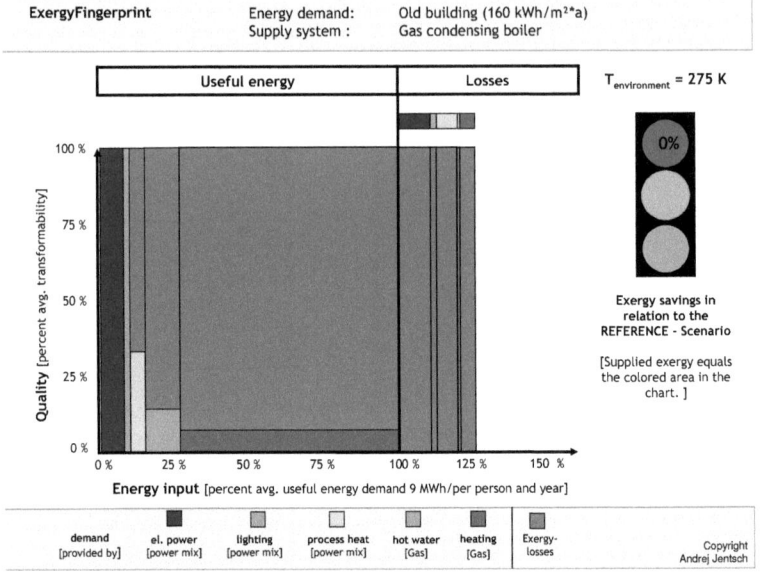

Figure 4.4: Reference scenario of the ExergyFingerprint of an old building supplied by the average german power mix and heat from a gas condensing boiler

The reference scenario is the basis for a comparative assessment using the ExergyFingerprint. The x-axis shows thermal, chemical and effective thermal transformation energy, which for applications above reference temperature equals energy. To simplify understanding, the axis has been simply labeled "energy", which is fully correct for thermal transformation energy above reference temperature and electrical transformation energy. The deviation of chemical transformation energy from the higher heating value is small, so that only a minor error is made for this type of transformation energy. Mechanical exergy has been neglected in the considered example. The scale on which energy is measured is a relative scale to allow easier comparison of different ExergyFingerprints. Knowing the real energy equivalent of $100\,\%$ as $9\,MWh/per\,person\,and\,year$, the absolute dimension of the energy demand is also known.

---

[8] After the ExergyFingerprint has been developed independently by Fraunhofer UMSICHT, it was found that a very basic first approach to the graphical characterization of exergy as a product of quality and quantity has been developed by Nieuwlaar and Dijk (1993). However, their presentation and the theory on which their presentation is based differ in most aspects from the one developed, so that the development of the ExergyFingerprint as it is, can be considered an original work of Fraunhofer UMSICHT.

4 Application of the transformability evaluation and analysis method

Complementing the transformation energy on the x-axis, the average (avg.) transformability is plotted on the y-axis. The axis has been labelled „Quality" to simplify understanding of the meaning associated with transformability. The average transformabilities for the different types of useful energy have been calculated using the equations from Table A.2 on page 130 and data from Table A.31 on page 177.

The ExergyFingerprint thus allows to graphically express exergy as a product of quality and quantity. This can be considered a novel approach to understanding exergy basics without dealing with equations or thermodynamic theory.

To illustrate the way in which the ExergyFingerprint can help to understand results of thermodynamic comparative assessment, two further ExergyFingerprints have been added. Figure 4.5 shows the ExergyFingerprint of an old building that is supplied with heat from a block CHP plant by district heating, which is discussed in section 4.2.

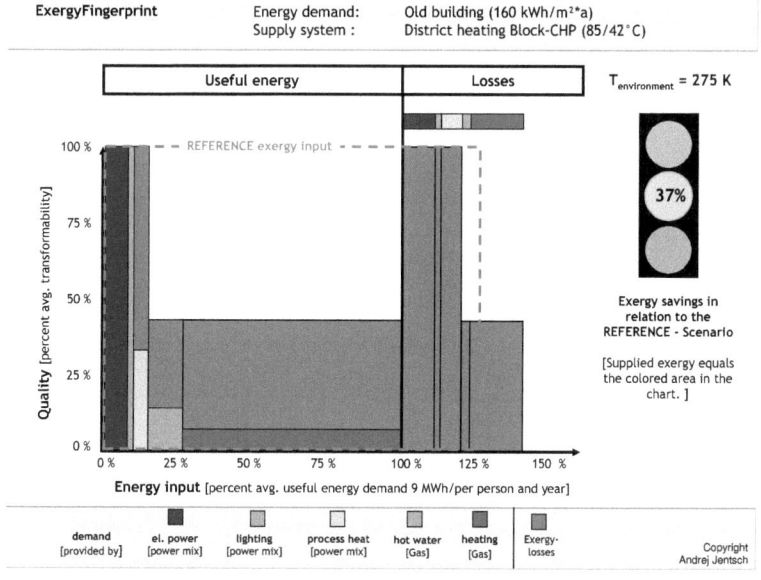

Figure 4.5: ExergyFingerprint of an old building supplied by the average German power mix and heat from a block heat and power plant.

To allow the characterization of further scenarios, the average annual heat demand for an old building had to be assumed. A value of $160\,kWh/m^2 \cdot a$ was found to be fairly realistic. Another possible scenario is presented in figure 4.6.

4 Application of the transformability evaluation and analysis method

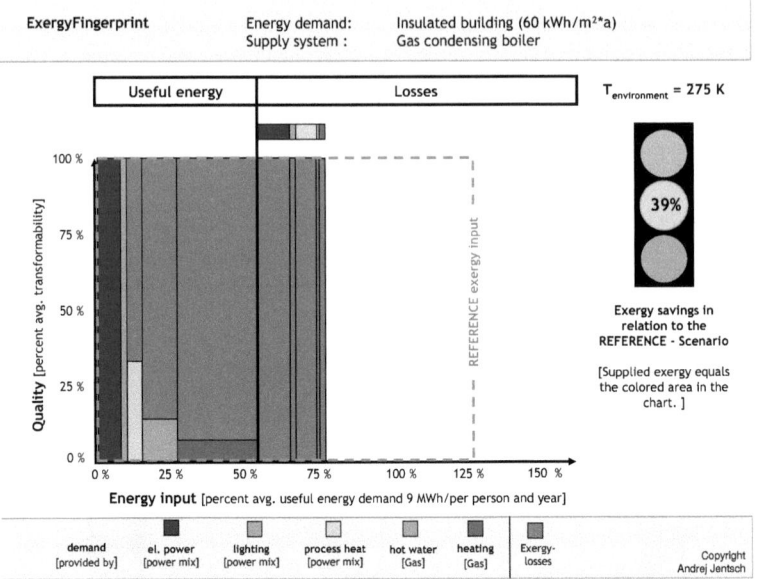

Figure 4.6: ExergyFingerprint of an insulated building supplied by the average German power mix and heat from a gas condensing boiler.

Comparing Figures 4.5 and 4.6 with Figure 4.4, the advantages of a separate evaluation of the qualitative aspects and the quantitative aspects of exergy become obvious. While it is usually diffcult to explain exergy to nonspecialists, a separation of exergy into a relative measure for quality - transformability - and a familiar technical property - energy - in a graphical way can simplify the communication of the exergy concept. In addition to the more transparent transformability and transformation energy assessment, a traffic light on the right hand side of the diagram summarizes the exergetic improvement of the considered scenario over the reference scenario, thus allowing a simple one dimensional ranking of supply / demand scenarios on a strict thermodynamic basis.

The ExergyFingerprint can help to make evaluation results of transformability assessment easily understandable and provide a novel perspective on exergy evaluation. Additionally, it helps to clarify that technologies which decrease the average transformability of the fuel for low transformability products can have an equally significant impact on saving exergy of primary energy as decreasing (transformation) energy demand.

4.7 Defining the term "LowEx" by using the transformability concept

Originally, the term LowEx has been introduced as an abbreviation for "low exergy". Several projects (VTT, 2003; lowex.info, 2009; LowEx.net, 2009) focus on the research of LowEx technologies, which

4 Application of the transformability evaluation and analysis method

are supposed to make good use of the "potential" of energy by utilizing low temperature heating and high temperature cooling. However, up to date quantified criteria which a technology has to fulfill in order to count as a LowEx technology have not been defined. Commonly, a low energy consumption is considered a precondition of low exergy systems (VTT, 2003, p. 122). This appears sensible since exergy includes qualitative and quantitative aspects. Despite this precondition, low exergy systems are yet defined independently of efficiency or specific consumption considerations. In Annex 37 (of the International Energy Agency) LowEx systems are defined as heating or cooling systems that allow the use of low valued energy as the primary fuel source (VTT, 2003). Using the exergy concept alone, it is difficult to exactly specify such LowEx systems, since an exergy value is not only influenced by its specific value but also by the size of the flows considered. The transformability concept can help to overcome this difficulty in defining LowEx systems. Using the average product transformability as an indicator for a system requiring a "low value" energy such systems can be exactly specified. Additionally, it appears sensible to define a minimum exergetic efficiency of the supply system and a maximum (transformation) energy demand of the supply target for such LowEx energy systems.

For the considered thermal supply systems, a possible suggestion for the limits of LowEx systems would be a maximum product transformability of 10 %, which is approximately the thermal transformability of hot tab water at 60 °C and a reference temperature of 3,5 °C. The exergetic efficiency of the supply system and the maximum (transformation) energy demand of the supply target could be defined in relation to a reference supply scenario and a desired improvement. E.g. if the exergy consumption should be reduced by 80 %, the maximum transformation energy demand of a LowEx system could be defined as half of the reference demand, while the minimum exergetic efficiency of a LowEx supply system ε_{LE}^{mn} could be defined as a function of the exergetic efficiency of a reference system ε_{RF}:

$$\varepsilon_{LE}^{mn} = \varepsilon_{RF} \cdot 2,5$$

According to this suggestion, the heat supply systems considered in section 4.2 on page 67 could potentially be LowEx heat supply systems, since the required product transformability is approximately 7 %. If assuming the gas condensing boiler system as the reference supply system, the minimum exergetic efficiency of a LowEx system would be 15,9 %, thus identifying the geothermal heat supply system as the only LowEx supply system of the alternatives considered and according to the LowEx definition suggested here. Since the supply target has been defined equal for all systems, it is not possible to identify a LowEx supply target from the considered examples.

The extension of the LowEx term to components as suggested in lowex.info (2009); LowEx.net (2009) and enob.info (2009) can be considered problematic, as those components, if not integrated into a LowEx system, do not necessarily improve exergetic performance. It is therefore proposed to consider the term "LowEx-ready" for such components which concerning their fuel and product flows fulfill the same demands as LowEx supply systems, i.e. a maximum average transformability of the required product and a higher exergetic efficiency in relation to the reference technology. The balance boundaries of "LowEx" systems that integrate such "LowEx-ready" components could be defined in accordance with the rules for comparative energy system evaluation presented in section 3.3 ff.

While the exact values of the limits that define a LowEx system have to be discussed in a circle of experts, the LowEx-term can already be communicated more transparently using the transformability concept and method. LowEx energy systems could be described as systems with an energy demand of low quality (transformability) which are a combination of exergetically efficient supply systems and a supply target with a comparably low energy demand. The use of the LowEx-label can thus provide a means to promote the use of heat sources with low transformability, such as waste heat, geothermal heat or heat from solarthermal collectors or of direct cooling systems such as direct seawater cooling.

5 Discussion and Conclusion

5.1 Transformation energy and transformability

The separation of the quantitative and the qualitative aspects of exergy provides a new perspective on comprehensive thermodynamic evaluation of mass and energy transfers. This allows a transparent characterization of processes and flows, which is as universally applicable as evaluation using exergy. The separate evaluation of quality and quantity makes the differences between considered flows apparent, while at the same time changing the point of view from the total value of a flow that can be obtained using exergy to a more differentiated perspective which can be used to complement exergy.

It can be argued that the transformability concept provides the basis for a new simplified understanding of exergy as a product of thermodynamic quality and (transformation) energy. The effects of the first law are implicitly present in transformation energy which allows the association of a characteristic energetic value with any flow. It can be viewed as a pragmatic way to increase the options for energy-based evaluation, allowing a principal quantitative comparison of mass flows and energy transfers alike, even if mass flows are characterized mainly by a deviation in specific entropy rather than in specific enthalpy from reference conditions, like e.g. pressurized gas flows at reference temperature. Since energy is a property with which a large part of society is familiar, the possibility of considering transformation energy as a form of energy could provide a scientific foundation for the common belief that all useful flows "have an energy". Additionally, this quality-independent property allows a clear and universal assessment of "size" associated with a flow. Energy and mass are limited to some types of transfers, while exergy includes quality aspects, which can result in a distortion of its ability to indicate quantitative effects. Thus, neither the consideration of mass nor of energy or exergy can provide such a direct, universal and unambiguous measure of quantity as transformation energy.

However, the new concept also provides ground for controversial discussion. The use of this universal energetic assessment parameter requires a significant increase in computational effort for many types of exergy flows. That makes it questionable whether this concept is a useful option for manual calculation. Additionally, in order to satisfy the first law of thermodynamics, the association of more than one value with a given transfer is required. The use of compensation heat flows significantly increases the complexity of the theory and makes it more difficult to gain a deeper understanding of the concept.[1] Therefore, the concept should be considered as an extension but not as a replacement or competition for the exergy concept.

The second law aspects of exergy, i.e. the influence of exergy destruction on exergy analysis, can be considered to be represented by transformability. The expression of the entropic aspects of a flow in a dimensionless property, which can only have values between 0% and 100%, is very convenient for the communication of some implications of the second law on technically relevant transfers. A conclusion from the second law of thermodynamics can be expressed in terms of transformability as:

---

[1] Therefore, this aspect is discussed separately in subsection 5.2.1.

5 Discussion and Conclusion

The average transformability (quality) of the transformation energy output of a process can never exceed the average transformability of its input.

It is important to take into account that the direction of input and output is defined based on the direction of the transformation energy flows. Negative signs can consistently be interpreted as indicators that the direction of a transformation energy or exergy flow is opposed to the mass or energy transfer these flows are associated with. If this precondition is accepted, the transformability concept allows a new and consistent perspective on some effects of the second law of thermodynamics that could make it easier to understand important aspects of its impact in engineering.

Transformation energy and transformability contain all information that is contained in exergy. The difference between the transformability concept and the exergy concept lies within the different perspective on the considered flow. The exergetic view can be seen as a top-down view on the potential to do work that is associated with a flow in a given environment. It assesses the "total value" associated with a flow not discriminating between quantitative and qualitative influences. The transformability concept focuses the perspective of the relative ability to do work and provides insight into the "specific value" that can be associated with the flow as well as assessing its "size" using transformation energy. As a complementary concept the transformability theory could be useful in explaining and quantifying the qualitative and quantitative aspects of exergy.

Concluding, it can be said that the transformability concept provides a strictly scientific perspective on the evaluation of transfers, which provides a differentiation concerning the size and the quality associated with the flow. It shows that only transfers that can be associated with a temperature difference to the environment are theoretically limited in their transformability and are therefore of lesser quality. Based on the definition of transformation energy, a consistent concept has been developed that proves and allows to communicate that theoretically all types of nonthermal transfers are of equal "specific value".

5.2 Transformability assessment and analysis

5.2.1 The transformation energy balance and the compensation heat flow

The transformation energy balance can require significantly more effort than an exergy or energy balance, which has been demonstrated in the figures in appendix A11 on page 152 ff. The major contribution to this increased complexity is caused by the association of some mass- or energy transfers not only with a valuable transformation energy but also with a so called compensation heat flow. It has been introduced as a consequence of the assumption that the transformation energy balance has to fulfill the law of conservation while at the same time all transformation energy flows have the same algebraic sign as the corresponding exergy transfers. It is notable that certain compensation heat flows can be very large. Although the absolute value is of little importance to evaluation, since following the explanation in subsection 3.2.3 on page 54 ff. only effective compensation heat flows are considered in the transformability evaluation, it is still interesting to find an interpretation for their meaning.

5 Discussion and Conclusion

A transformation energy flow that is associated with a mass or energy transfer is an indicator for the amount of energy minimally required to transform the exergy associated with the considered transfer into work. If also a compensation heat flow is associated with this transfer, this can be seen as an indicator of the energetic contribution of the environment to the mechanical or nonreactive transformation energy of that flow. For these two types of transformation energy the compensation heat flow has an absolute value close to the absolute value of the considered type of transformation energy associated with the transfer, indicating that the energy to generate work from pressure and concentration differences is provided mainly by the environment.

The absolute value of a thermal compensation heat flow for heat flows at temperatures below reference temperature is always larger than the matching transformation energy, so it cannot be interpreted as the contribution of the environment to the considered transformation energy flow. It equals the sum of the absolute values of the heat input and the heat output of a reversible heat engine generating an amount of work equal to the exergy associated with the considered flow. Thus, it compensates the attribution of transformation energy and exergy to an energy flow not directly responsible for work generation, since the energy input that allows to generate work, if considering heat flows at temperatures below reference temperature, is obtained from the environment. This unintuitive attribution of the ability to work to the potential discharge flow is a consequence of defining the heat from the environment as being without value, since it is infinitely available. This central aspect of exergy theory can be perceived as being problematic for a correct understanding of exergy in a general thermodynamic context.

An indicator for the difficulty to understand the exergy concept correctly is the long popularity of the anergy concept. Anergy was intended to complement the concept of useful energy or exergy with a value to describe „useless" energy at reference conditions. Only in the last two decades the anergy concept which has been popular in Europe is increasingly discredited, since it has been identified as not being sufficiently universal (Bosnjakovic and Knoche, 1998; Szargut, 2005).

The difficulty that results from the definition of heat at reference temperature as "worthless", and as a consequence the fact that a straightforward interpretation of thermal compensation heat flows is not possible, has therefore to be accepted as a side effect of the exergy definition. It illustrates the problem resulting from the definition of a reference state with a quality value of zero and of the association of the work potential of the combined system with the flow, independently of the fact whether it provides the energy to generate work or not.

5.2.2 Interpretation of the results of transformability evaluation and analysis

It has been shown by the examples in section 4.3 on page 72 ff. that various energy supply systems with different supply targets and at different supply temperatures can be evaluated using comparative transformability evaluation. Since the results are in principle on the same scale, they can all be subjected to a very general type of assessment.

As a result of the assessment based on Figures 4.2 on page 77 and 4.1 on page 75, several effects can be noticed. Firstly, the reference temperature has a very strong influence on transformability ratio for

5 Discussion and Conclusion

supply systems providing low-transformability heat. This is universal to heating and cooling systems, so that an accurate comparative transformability assessment of such supply systems requires accurate environmental temperature data for the period of system operation. The strong dependence of the transformability ratio on the reference temperature is an effect of the low product transformability, which is caused by temperatures comparably close to reference temperature. Small changes in reference temperature therefore significantly impact the transformability of the product.

The considered exemplary heating and cooling systems vary in their transformation energy efficiencies from 49% to 95%. This indicates significant external improvement potential, especially for heat pumps and compression refrigeration machines. Since the consideration of the relevant flow charts shows mainly losses at the power plants providing electricity, an improvement of the electric efficiency of the power generators, e.g. by using the cross-comparable section of wind power generators, would significantly improve the external sophistication of these heating systems. Another theoretical option for increasing transformation energy efficiency of heat pumps is the use of expanders that help to drive the compressors instead of the use of throttling valves to decrease the pressure after the working fluid exits from the condensation state[2]. In such way, less fuel would be required to drive the processes and lower losses from electricity generation and the heat pump or refrigeration machine would decrease transformation energy efficiency.

While the transformation energy efficiencies are mostly above 50%, the transformability ratios for the considered systems are all below 50%. This indicates that all analysed processes suffer mainly from a suboptimal use of the average fuel transformability. The comparably low transformability ratio indicates that most considered processes for the supply of a low-transformability product are not very well suited to the considered task from a thermodynamic point of view. A large part of this suboptimal suitability can be explained with the low transformability of the required product and the necessity to have a temperature difference in order to exchange heat with a limited heat exchanger area. However, the considered compression heat pump has a comparably high maximum transformability ratio of 60% , which can be interpreted as an indicator of a good suitability of the heat pump principle for the supply of room heat. The large difference between the real and the maximum transformability ratio of the compression refrigeration machine system and heat pump system in conjunction with the low transformation energy efficiency indicates the need for improvements of these heat supply systems that lead to a decrease in combustible fuel consumption. A large contribution to the improvement of the transformability ratio of these systems can be expected from the measures taken to improve transformation energy efficiency.

The maximum transformability ratio of the boiler system is very low, thus identifying it as a system that is poorly suited for the use of combustible fuels, even in an energetically ideal case. This system should be therefore replaced by a more suitable one wherever possible. The use of the two novel ratios can help to communicate the necessity of such an exchange, as it relates to the familiar and comparably high energy efficiency of boiler heat supply systems but adds a ratio that quantifies its insufficient suitability for the considered supply task. Since the transformabilty ratio can be communicated as a

---

[2]However, in practice the replacement of throttles by expanders is challenging, since the replacing expander would have to be able to handle changes in aggregate state and eventually would have to allow heat transfers to the fluid in order to provide an exit flow at the same parameters as the throttle. (C. Pollerberg 2009, personal communication, 18 August)

5 Discussion and Conclusion

ratio of qualities, it does not require the introduction of fundamentally new concepts like the exergy concept to explain this consequence of thermodynamic analysis.

It has been noticed that the transformability ratio is decreasing for all technologies the closer the reference temperature is to the target supply temperature. A general conclusion from this could be to avoid the use of heating and cooling supply technologies as long as the temperature difference between real temperature within a supply volume and the the target supply temperature remains tolerable.

In the following, the results of the comparative assessment of heating and cooling systems are discussed. Of the considered heating systems, the geothermal district heating system and of the considered cooling systems, the direct seawater cooling system have the highest transformability ratios and the highest exergetic efficiencies. It can be concluded that in order to provide thermal transformation energy, it is best to use mainly a natural source of thermal transformation energy with an average fuel transformability that is as close as possible to the required transformability. The criterion of transformability ratio is therefore significantly more relevant when choosing a thermal supply system, than the transformation energy efficiency, which for the case of heating systems equals the conventional energy efficiency. The benefit of the transformability assessment lies in its transparency and the possibility to communicate the results of exergy-based analyses differently. Instead of having to introduce a new property (exergy), the transformability ratio can be introduced as a ratio of product and fuel qualities and provide a measure of an intuitively understandable measure of process suitability.

The application of the transformability concept to process analysis in sections 4.4 on page 78 ff. and 4.5 on page 84 ff. shows that the largest impacts of thermodynamic inefficiencies are not found in the area of external improvement potentials but in their influence on average fuel quality or product quality. All technologies that use thermal transformation energy indirectly through heat exchangers, suffer more from a low match of product and fuel quality than from actual transformation energy losses, independently whether pressure decreases or heat losses are the problem. Only expanders and compressors show a larger decrease in transformation energy efficiency than in transformability ratio for a given heat loss, which is mainly owed to the high transformability of the major product of $100\,\%$. This implies that the largest optimization potential of most processes with thermal products lies in parameter and design optimization for the considered processes.

The transformability analysis also helps to analyze the vapor-cascade refrigeration machine from section 4.5 on page 84 ff. in greater detail than by an evaluation using exergetic efficiency alone. It becomes evident that none of the considered components but the the process as a whole, due to the presence of dissipative components (the the cooler), has a large potential for external optimization.

All of the discussed examples indicate that a consistent and universal analysis of processes using transformability analysis is possible and could prove sensible as an extension of the established exergy analysis. Exergetic efficiency allows an evaluation of the total performance of the considered systems, thus remaining one of the best thermodynamic criteria for the evaluation of actual process choices. The use of the transformability ratio and transformation energy analysis on the other hand can help to better understand the fundamentally different optimization potentials of a process. On the

5 Discussion and Conclusion

one hand a process can be optimized by decreasing avoidable external losses, e.g. by using waste transformation energy flows, while on the other hand a process requires internal optimization, which could be expressed as a need to take measures which bring the average transformability of the fuel closer to the average transformability of the required product. The use of the transformability concept in thermodynamic analysis could thus help to increase the popularity of exergy-based evaluation, by allowing a new way of communication of exergetic optimization potentials. The ExergyFingerprint for example allows a new perspective on supply scenarios which is exergy-based, but does not require familiarity with the exergy concept, as the exergy concept and its difference to the energy concept can be explained using the ExergyFingerprint itself. Furthermore, the use of a two dimensional perspective on exergetic efficiency like in Figures 4.1 on page 75 or 4.2 on page 77 helps to identify various technology evaluation results despite changes in reference temperature, and allows to understand exergetic efficiency of thermal supply systems as consisting of a part nearly independent and a part dependent on reference temperature. However, due to the high complexity of the method, calculations should be automated and the presentation should explain the results to people not professionally occupied with thermodynamics in a simplified way using the words "energy" and "quality" instead of transformation energy and transformability.

Concluding, it can be stated that the transformability concept can lay the basis for better communication of the exergy concept. Flows that formerly had to be labelled as low-exergetic, can now be labelled more accurately as being low-transformability or simply low-quality flows. Additionally, the transformability ratio can be used as a direct and scientifically sound way to distinguish processes with large specific irreversibilities from systems with comparably low specific irreversibilities. The use of the transformability ratio can substitute evaluation methods which require a direct consideration of the property entropy, which is comparably difficult to understand, or the use of an exergy destruction based ratio, the value of which is dependent on the definition of the system boundary at which losses are evaluated. Thus, the developed definition of thermodynamic quality allows to describe one important task of engineering as improving the match between the average transformability demand and the average fuel transformability.

5.3 Advantages and Disadvantages of the transformability concept and the transformability evaluation and analysis method

The advantages and disadvantages of the transformability concept and the transformability evaluation and analysis method are discussed in this section to clarify the usefulness of the new concept and method.

5.3.1 Advantages of the transformability concept

The transformability concept allows a more transparent characterization of energy and mass transfers than the exergy concept.

Transformation energy and transformability allow a more transparent characterization of mass and energy transfers by making it possible to intuitively assess the „size" and the specific thermodynamic

5 Discussion and Conclusion

value of a flow. While exergy provides a measure of total thermodynamic value of a flow, the origin of this value becomes clearer by also using the transformability concept, which allows to clearly distinguish large low-quality flows from smaller high-quality transfers, which in principle could be associated with the same exergy value.

The transformability concept provides a new perspective on exergy which allows better communication.

Frequently misunderstandings result when communicating and explaining exergy. Some of them are summarized in appendix A1 on page 123 ff. Some of these misunderstanding can be avoided if using the transformability concept. Using a simplified version of the transformability concept, exergy can be explained as a product of energy and quality. Emphasizing the importance of the quality aspect for all processes exergy can be clearly distinguished from energy and explained as being more universal. The theory behind the transformability concept is derived based on reversible processes. This rather practical derivation clearly demonstrates that exergy is always a property of the combined system of flow under consideration and the reference environment. Therefore, exergy is only associated with a considered flow but not a property of it.

The transformability concept allows to quantify the specific value of heat and thermal energy and proves the theoretical equivalence of all other types of transfer concerning their theoretical transformability.

Using exergy, it is obvious that thermal, conductive and effective thermal exergy flows are always associated with significantly larger energy transfers. Thus, heat especially if at temperatures close to reference temperature is considered a low-value form of energy. However, the quality of all other types of useful transfers, such as of mechanical exergy, chemical exergy of different fuels or nonreactive exergy is significantly more difficult to assess, since in some cases no or little enthalpy transfer is associated with these flows. With transformability a property has become available that allows to assess the special role of heat transfers in the context of all other types of energy and mass transfers. Only flows associated with exergy due to a temperature difference from the environment have a transformabilty lower than 100 %. This means that essentially all other types of transfers are equally valuable in regard to their thermodynamic value. Since the transformability concept is a strictly thermodynamic concept, it provides an assessment of "quality" independent of subjective choices, if the reference environment is modelled according to the real environment. With quality measured in "percent transformability", a property has become available to point out the special characteristic of heat and thermal energy and to quantify it.

The transformability concept is helpful when defining the requirements for the label "LowEx"

While exergy as a product of quality and quantity makes it difficult to specify the term "LowEx", the product transformability can be helpful when defining this term. Supply systems and components that are "LowEx" have a required product that can be characterized by a maximum product transformability.

5.3.2 Advantages of transformability assessment and analysis

The transformability analysis and assessment method allows to evaluate processes concerning their thermal sophistication and their suitability with respect to the given supply task, thus giving an

5 Discussion and Conclusion

indication of the primary area of improvement.

With the definition of the transformation energy efficiency and the transformability ratio, two evaluation ratios have been found that provide a universal assessment of the degree of external sophistication (transformation energy efficiency) and the degree of process suitability (transformability ratio). Improving transformation energy efficiency can in many cases be achieved up to 100 % by measures that do neither influence intensive process parameters nor process design such as the application of better insulation, sealing of leaks or the use of more sophisticated components, which use the same inputs and provide at least the same products as those components they replace. Thus, in contrast to exergetic efficiency, which does not indicate whether the improvement potential is mainly internal or external, the evaluation using transformation energy efficiency provides a first general indication of the degree, to which a process reaches an external optimum.

The transformability ratio on the other hand gives an indication of process suitability, relating average product quality to average fuel quality. It allows to assess how well a considered process uses the provided specific potential of the fuel. A comparison of technologies based on the maximum transformability ratio helps to identify suitable process types even if their degree of external sophistication (transformation energy efficiency) is low.

Since the product of transformation energy efficiency and transformability ratio is exergetic efficiency, both ratios are equally important for total process sophistication. Thus, the ratio of the two which shows the lower value always indicates whether an optimization should focus on the decrease of external losses, i.e. low transformation energy efficiency, or on the improvement of process design and parameters, i.e. low transformability ratio.

To assess the degree of suitability of the process design, instead of the transformability ratio, which allows the assessment of the suitability of the process with all its losses, the maximum transformability ratio can be calculated. This evaluation ratio is calculated based on theoretically optimal values for the magnitude of the fuel flows, which means that the transformation energy efficiency of all components is set to 100% and unnecessary pressure drops are avoided. The intensive properties of all flows and the process design equal the one of the real process. The calculated value equals the maximum exergetic efficiency and can therefore be interpreted either as a degree of suitability of the process design or as the maximum thermodynamic sophistication achievable with this design. E.g. the use of this evaluation ratio can clarify and allow to quantify how badly suited boiler systems are in general to provide room heat.

The results of transformability analysis can be communicated in a novel way to people not professionally occupied with thermodynamics, since its basic properties relate directly to the familiar concepts of energy and quality.

Transformation energy efficiency can be communicated to people not professionally occupied with thermodynamics as advanced energy efficiency, since it allows an energy-efficiency-like assessment of most types of processes. It indicates the sophistication of the process in regard to external performance, which can frequently be improved by using components with lower losses, by applying external means such as insulation or leak sealing or by using waste flows. Since the concept of energy efficiency is essentially understood, a simplifying label of transformation energy efficiency as

5 Discussion and Conclusion

(universal) energy efficiency can provide a means to extend the known concept of energy efficiency to most technologies.

Transformability ratio, on the other hand, indicates the suitability of the process as a „consumer" of the fuel provided or as a supplier of the product, thus providing an opportunity to relable the problem of exergy destruction as a problem of quality destruction, thus avoiding the need to introduce the exergy concept directly while at the same time pointing out its central benefits over the energy concept. A process with a low transformabilty ratio is usually a poor choice for the application of the fuel it uses and should either be redesigned or replaced with a more appropriate system. Transformability ratio can thus be communicated as an indication whether the process used for a given task requires process parameter optimization. The achievable suitability of a given process can be assessed if considering maximum transformability ratio.

Based on the transformability evaluation the ExergyFingerprint as a graphical assessment tool for energy supply scenarios has been developed, which allows to clarify the difference between technologies that have similar transformation energy efficiency but fundamentally different transformability ratios. This assessment tool has been received with significant interest by government officials and engineers in the field of heat supply systems.

Transformability analysis can complement exergy analysis to obtain a higher transparency.

Transformability analysis allows to characterize system components and systems more transparently than an exclusively exergetic analysis, which provides an aggregated evaluation of quantitative and qualitative effects. Additionally, the dependence of exergy destruction and loss on the boundary at which lost heat flows are evaluated, makes it difficult to identify the external optimization potential of components. E.g. the transformability analysis of the considered vapor-compression cascade refrigeration machine shows clearly that the considered components are not causing the low transformation energy efficiency of the whole process. Based on exergetic efficiency, the impact of the cooler on fuel consumption would not be as directly visible as it is by using transformation energy efficiency.

5.3.3 Disadvantages

The transformability method is highly complex.

Although the transformability evaluation and analysis method is consistent, a full understanding of this theory is difficult, since it adds a new level of complexity to exergy-based evaluation. It appears therefore improbable that a deep understanding of the method will be gained by a large number of people. However, the occupation with the transformability method can help to understand exergy in a novel way thus improving the quality of the application of exergetic analysis and evaluation and the communication of its results.

The transformability concept and consequently transformability evaluation and analysis require an increased computational effort in comparison to exergy-based calculations.

The easiest way of calculating values for transformation energy efficiency and transformability ratio is to calculate exergetic efficiency and transformation energy efficiency and divide both ratios by one another. However, since only thermal transformation energy and transformation energy associated with conductive heat flows are calculated differently from exergy, only some flows and the compensation heat flows have to be computed additionally. Due to modern information technology and modern software, the calculation can be automated. The variety of definitions for compensation heat flows and transformation energy definitions is essentially requiring implementation in a dedicated software. As long as such a program is not widely available, the increased computational effort will limit the application of the transformability concept and the transformability assessment and analysis method.

Outlook

The transformability concept and the transformability assessment and analysis method can be the starting point for numerous research and development activities. It appears interesting to investigate the possibility of extending advanced concepts which use exergy to be based on transformation energy and transformability. For example, it could be attempted to substitute exergy with transformation energy and transformability in the hierarchically structured exergy analysis method presented by Hebecker et al. (2004) and the concept of advanced exergy analysis presented by Tsatsaronis and Park (2002) and Morosuk and Tsatsaronis (2008). Potentially, this could provide a greater transparency to process analysis, in situations which require clarification in regard to a separate evaluation of quanitative and qualitative effects . However, such an improved transparency could come only at the expense of even greater complexity. Thus, a further development of the transformability analysis appears only justified when the claims of improved communicability and transparency have been confirmed by practicing engineers.

However, the next step to take appears to be the extension of ExergyFingerprints to cooling technologies and to other technologies with effective compensation heat flows. Furthermore, it can be investigated whether the presentation can be switched to presenting actual transformabilites instead of average values to provide experts that can deal with such a decreased level of aggregation with a more accurate graphical representation of thermodynamic supply scenario characteristics.

Another challenging area in which the application of the novel assessment method could be attempted is its integration into exergoeconomic analysis. It could be possible that a relation of cost factors to transformability (specific cost / quality) or a relation of cost factors to transformation energy (specific costs / per indestructible quantity) would improve the transparency of exergoeconomic evaluation where necessary. On the other hand, transformability ratio and transformation energy efficiency could potentially complement exergetic efficiency in multidimensional technology assessments, like the one suggested by Radgen and Oberschmidt (2006).

Outlook

Apart from continuing research concerning transformability evaluation and analysis, its complex computation requires a certain degree of automation if it should find a wider area of application. A program for exergy analysis could therefore be extended to include transformability and transformation energy calculation basics, so that by increasing the utilization of the new properties, more about the benefits and problems of their application can be learned. As a long term goal, a comprehensive software suite that includes property data, standard chemical exergies and calculation routines for exergy, transformation energy and transformability as well as definitions of exergetic and transformation energy efficiencies for a broad variety of common processes could be developed. Additionally, it could allow the automatic generation of graphical assessment tools such as two dimensional transformation energy efficiency and transformability charts or ExergyFingerprints.

Summary

Exergy can in principle be considered a product of quantity and quality. Yet, until now no theory has been developed that allows a consistent separation of these two aspects of exergy. One basic goal of this work was therefore to find and derive definitions that allow a separate assessment of exergy-based quality and quantity aspects. This separation appeared necessary to improve communicability of the exergy concept and of the results of exergy analysis. With the definition of transformation energy, a measure of the quantitative aspect of exergy has been given a scientific foundation:

A type of transformation energy is the amount of energy input into a reversible process required to transform the matching type of exergy (e.g. thermal, mechanical...) completely into work. The only energy exchange between environment and the considered flow is the exchange of heat at reference temperature.

The capacity of doing work has been accepted as a measure of thermodynamic quality, which for heat flows above reference temperature can be defined as the ratio of exergy associated with the heat flow to energy. However, it was shown that this approach is limited in its applicability, neither can it be used to sensibly assess the "quality" associated with heat flows below reference temperature nor is it suitable for the evaluation of transfers associated with mechanical or nonreactive exergy. Thus, a novel measure of thermodynamic quality associated with a flow has been defined, which has been termed transformability. It is a relative measure indicating the share of the considered transformation energy that can be transformed into work. Its value is always between 0% and 100%. It thus allows an assessment of the quality associated with a combined system of considered flow and environment on a dimensionless and familiar scale.

Based on the concept of transformation energy and transformability, which can be referred to as the "transformability concept", evaluation ratios have been defined which can be interpreted as a degree of external sophistication of a technology and a degree of process suitability. The degree of external sophistication has been termed transformation energy efficiency, while the degree of process suitability was given the name transformability ratio. The product of both is the exergetic efficiency.

Using the newly introduced transformation energy balance, a structured method for the comparative assessment of energy supply technologies has been developed that is used to demonstrate the application of the novel ratios. This method uses generalized rules for the definition of balance boundaries to ensure cross-technology comparability. Firstly, all chosen supply systems are considered to fulfill the same exergetic demand with a common specification. A generic home has been chosen as the supply target for the comparison of domestic heating and cooling supply systems which have to supply or extract heat to maintain an indoor temperature of $295\,K$. Secondly, the supply systems are separated into cross-comparable and technology-specific subsystems. The cross-comparable subsystems are defined as those parts of the total supply system, into which storable primary energy enters or into which the first storable secondary energy enters if processes using nonstorable primary energy such as solar thermal heating system are evaluated. Thus, the boundary on the supply side is set directly before or directly after the first energy converter in the process chain. The specific characteristics of the transformation technologies that allow the conversion of nonstorable primary energy to storable

Summary

energy forms, can be separately evaluated within the respective technology-specific subsystems. The application of the storability criterion is intended to ensure a „just" comparison of different technologies, since only storable forms of exergy can be utilized on demand.

In addition to a structured procedure for the definition of evaluation boundaries, a consistent exergy-based approach to the attribution of fuel to heat from combined heat and power processes has been developed. This method results in the same attribution factor that has been previously used for the exergy-based allocation of carbon dioxide emissions from combined heat and power and which has recently been recommended for an ecological attribution of fuel to heat from combined heat and power.

Using transformation energy efficiency and transformability ratio and following the rules for boundary definition and evaluation of combined heat and power, seven thermal energy supply systems have been assessed and discussed. It can be shown that the improvement potential of all considered technologies lies mainly in improving system parameters or process design and not as much in further mitigating energy losses. Among all technologies, those heating and cooling technologies perform best that fulfill the thermal supply task using mainly thermal transformation energy of low transformability. Furthermore, it can be shown that heat from a block heat and power plant can be similarly well suited to the supply of heat at room temperature as a generic ground-source heat pump, while providing heat at a higher average temperature.

Finally, the use of the transformability concept for the purpose of process analysis is demonstrated based on a discussion of various basic processes and the analysis of a vapor-compression cascade refrigeration machine. The analysis of these processes, using transformation energy efficiency and transformability ratio in addition to exergetic efficiency, provides a more differentiated perspective on the operation of the analyzed processes than the assessment with exergetic efficiency alone. Similarly to the results of the comparative evaluation, the use of the novel method can show that the major optimization potential of the considered processes usually lies in an improved process design or parameter optimization. Further increasing transformation energy efficiency of these processes is usually difficult to achieve as it is already close to $100\,\%$. Thus, the transformability analysis can show that independent of the boundaries for the evaluation of heat losses, which otherwise play an important role in distinguishing exergy destruction from exergy losses, exergetic efficiency is usually stronger influenced by transformability destruction than by avoidable losses to the process surroundings.

A first practical application of the transformability assessment method has been found with the ExergyFingerprint, a graphical assessment and characterization tool for the evaluation of energy supply scenarios which has gained attention of engineers and government officials in Germany.

The newly introduced properties and ratios have some advantages over the use of the exergy concept alone.

The transformability concept allows a more transparent characterization of energy and mass transfers than the exergy concept. It provides a new perspective on exergy allowing to communicate it in a novel and eventually easier understandable way. It also allows to quantify the temperature-dependent quality of heat and thermal energy and proves the theoretical equivalence of all other types of transfer in regard

Summary

to their theoretical thermodynamic value. The definition of a maximum product transformability can help to define the label "LowEx".

Transformability evaluation and analysis allow to evaluate processes concerning their external sophistication and their suitability, thus giving an indication of the primary area of improvement. If simplified the results of a transformability analysis can be easier to communicate to people not professionally occupied with thermodynamics, isince its basic properties relate to the familiar concepts of energy and quality. Additionally, transformability analysis can help to characterize system components more clearly than by using exergetic efficiency alone.

The major disadvantage of the transformability concept and method is its more complicated calculation procedure, which limits the application of the concept as long as dedicated software tools are not available. Additionally, the theory has a level of complexity significantly exceeding that of the exergy concept, so that it is doubtful that a large number of people will gain a deeper understanding of the concept.

Concluding, the transformability concept and the transformabilty method can be considered useful contributions to thermodynamic theory. They provide a new, scientific and useful perspective at exergy, exergetic efficiency and supply system evaluation and lay the basis for a new perspective on the exergy concept and its difference from the familiar energy concept. Additionally, they allow to present the results obtained from thermodynamic analysis using innovative graphical evaluation tools, like the ExergyFingerprint.

Zusammenfassung

Exergie kann als das Produkt aus (energetischer) Quantität und (thermodynamischer) Qualität verstanden werden. Bisher existiert jedoch keine theoretische Grundlage auf Basis derer man eine wissenschaftlich fundierte Trennung dieser beiden Aspekte der Exergie vornehmen könnte. Ein grundlegendes Ziel dieser Arbeit war es daher, geeignete Definitionen zu finden und mit Hilfe dieser ein Konzept abzuleiten, welches eine schlüssige Definition von Exergie-basierter Quantität und der entsprechenden Qualität ermöglicht. Vor allem die Schwierigkeiten, Exergie zu kommunizieren, bilden die Motivation für die Entwicklung eines Konzeptes, welches Exergie in Anlehnung an bekannte Konzepte erklärt, ohne dabei zu Trugschlüssen zu verleiten. Mit der Definition der Wandlungsenergie wurde eine Größe gefunden, welche ähnlich universell ist wie die Exergie und mit deren Hilfe der quantitative Aspekt der Exergie bestimmbar wird.

Eine betrachteter Typ der Wandlungsenergie ist die Summe aller Energiezuflüsse in einen reversiblen Prozess, welche notwendig sind, um den entsprechenden Exergietyp (thermisch, mechanisch...) vollständig in Arbeit zu wandeln. Dabei findet zwischen der thermodynamischen Referenzumgebung und dem betrachteten Masse- oder Energiestrom ausschließlich ein Energieaustauch in Form von Wärmeströmen bei Umgebungstemperatur statt.

Die spezifische Fähigkeit, Arbeit zu verrichten ist anerkanntermaßen ein Maß für die thermodynamische Qualität. Für Wärmetransfers oberhalb der Referenztemperatur lässt sich diese Qualität aus dem Verhältnis von Exergie, welche mit einem Wärmestrom verbunden ist, zu dessen Energie berechnen. Es wird jedoch schnell offensichtlich, dass dieser Ansatz nur begrenzt anwendbar ist. Beispielsweise liefert das besagte Verhältnis keine sinnvollen Werte für die Qualität von Wärmeströmen unterhalb der Referenztemperatur sowie für die Qualität, welche mit Masseströmen mit einer Druckdifferenz zur Umgebung verbunden ist. Daher wurde die Wandelbarkeit als Verhältnis von Exergie zu Wandlungsenergie definiert. Sie gibt an, zu welchem Anteil die mit einem Transfer verbundene Wandlungsenergie in Arbeit transformierbar ist und hat aufgrund des Energieerhaltungssatzes immer Werte zwischen 0% und 100%. Damit wird die Bewertung der thermodynamischen Qualität auf einer ein universellen, dimensionsloses und gewohnten Skala ermöglicht.

Aufbauend auf dem Konzept der Wandlungsenergie und der Wandelbarkeit, welches im Folgenden als das Wandelbarkeitskonzept bezeichnet werden soll, wurden zwei Verhältnisse definiert, die Wandlungsenergieeffizienz und das Wandelbarkeitsverhältnis, welche als Grad der externen Güte sowie als Grad der Prozesseignung interpretiert werden können. Das Produkt der beiden neu eingeführten Größen ist die exergetische Effizienz.

Unter Verwendung der für die so genannte Wandelbarkeitsmethode entwickelten Wandlungsenergiebilanz wurde eine strukturierte Bewertungsmethode für Energieversorgungssysteme entwickelt. Diese Methode zeichnet sich insbesondere durch eine universal anwendbare und strukturierte Bilanzgrenzensetzung aus. Um die Vergleichbarkeit zu gewährleisten, müssen alle Energiesysteme das gleiche Produkt bereitstellen. Ein exemplarisches Haus wurde als Verbraucher gewählt, für welches sich die Versorgungsaufgabe stellt, die Raumtemperatur trotz Wärmeverlusten aufrecht zu halten. Weiterhin wurden die Energiesysteme in quervergleichbare und technologiespezifische Untersysteme zerteilt.

Zusammenfassung

Die quervergleichbaren Subsysteme sind so definiert, dass die Eingangsströme alle aus speicherbarer Primärenergie bestehen oder das Produkt der Wandlung einer nicht-speicherbaren Primärenergie in speicherbare Energie sind. Die technologiespezifischen Subsysteme, welche diese Wandlung bewerkstelligen, können anschließend zusätzlich bewertet werden. Die Speicherbarkeit erscheint als wesentliches Kriterium zur Differenzierung von Eingangsströmen, da nur speicherbare Energieträger bedarfsgerecht einsetzt werden können. Zusätzlich zur strukturierten Vorgehensweise bei der Festsetzung der Bilanzgrenzen für Technologievergleiche wurde ein schlüssiges Exergie-basiertes Vorgehen für die Zuordnung eines Brennstoffanteils zur Wärme aus Kraft-Wärme-Kopplung abgeleitet. Der sich daraus ergebende Attributionsfaktor entspricht dem Exergie-basierten Allokationsfaktor für Kohlendioxidemissionen aus Kraft-Wärme-Kopplung und einem aktuellen Vorschlag zur ökologischen Bewertung von Wärme aus Kraft-Wärme-Kopplung.

Unter Verwendung der dargestellten Bewertungsmethode wurden anschließend sieben einfache Energieversorgungssysteme bewertet und diskutiert. Anhand dieser Beispiele kann gezeigt werden, dass das Verbesserungspotenzial der meisten Systeme vor allem in der Verbesserung der Systemparameter und des Prozessdesigns und weniger in der direkten Verminderung oder Vermeidung externer Verluste liegt. Von allen verglichenen Technologien zur Heizung und Kühlung stellen sich diejenigen am besten dar, welche vor allem thermische Wandlungsenergie niedriger Wandelbarkeit für die Befriedigung von thermischen „Bedürfnissen" einsetzen. Weiterhin kann gezeigt werden, dass die Wärme aus Kraft-Wärme-Kopplung ähnlich gut für die untersuchte Anwendung geeignet sein kann wie Wärme, welche mittels elektrischer Kompressionswärmepumpen und Erdsonden bereitgestellt wird, obwohl eine höhere Temperatur und damit eine höhere Wandelbarkeit der Wärme dem Versorgungsziel „Haus" zur Verfügung gestellt wird.

In einem letzten Schritt wurde die Eignung der Wandelbarkeitsmethode zur thermodynamischen Analyse anhand von diversen Beispielen untersucht. Die Untersuchung zeigt, dass die zusätzliche Anwendung der Wandelbarkeitsanalyse auf die betrachteten Beispiele eine differenziertere Perspektive auf den Betrieb der analysierten Prozesse ermöglicht, als es mit Hilfe der exergetischen Effizienz allein möglich wäre. Auch hier zeigt sich deutlich, dass das vorrangige Optimierungspotenzial der verschiedenen Komponenten im Bereich verbesserten Prozessdesigns und der Parameteroptimierung liegt. Beispielsweise ist eine weitere Erhöhung der Wandlungsenergieeffizienz für viele Komponenten der untersuchten Dampf-Kaskaden-Kompressionskältemaschine nicht möglich, da sie ohnehin schon bei nahezu 100 % liegen. Mit Hilfe der Wandelbarkeitsanalyse kann unabhängig von der Definition der Bilanzgrenzen für die Bewertung von Verlustwärmeströmen, welche in der exergetischen Bewertung eine große Rolle spielen, gezeigt werden, dass die exergetische Effizienz von thermischen Versorgungssystemen vor allem durch Wandelbarkeitsvernichtung beeinflusst wird und nicht durch theoretisch vermeidbare Wandlungsenergieverluste.

Eine erste praktische Anwendung für die Wandelbarkeitsmethode wurde mit dem ExergyFingerprint gefunden, welcher die grafische Bewertung und Charakterisierung von Energieversorgungsszenarien erlaubt. Diese Entwicklung wurde von Politik und Praxis in Deutschland mit Interesse aufgenommen.

Zusammenfassend kann gesagt werden, dass die neu eingeführten Größen und Bewertungsverhältnisse einige Vorzüge haben, wenn sie ergänzend zur Exergieanalyse verwendet werden.

Zusammenfassung

Das Wandelbarkeitskonzept ermöglicht eine transparentere Charakterisierung von Energie- und Massetransfers als das Exergiekonzept. Es ermöglicht weiterhin eine neue konsistente Perspektive auf die Größe Exergie als Produkt von thermodynamischer Qualität und Energie und erlaubt somit eine neue Kommunikation derselben. Auch kann mit dem Wandelbarkeitskonzept die Abhängigkeit der Qualität thermischer Energie von der Temperatur quantifiziert und die theoretische Gleichwertigkeit aller nicht-thermischen Transferarten belegt werden. Mit Hilfe der Definition einer maximal benötigten Wandelbarkeit kann zusätzlich eine Grundlage zur genauen Definition des bisher schwer fassbaren „Niedrig-Exergie" Begriffs geschaffen werden.

Die Wandelbarkeitsmethode ermöglicht es zusätzlich, Prozesse hinsichtlich ihrer externen Güte und Ihrer Eignung zu bewerten und gibt somit Aufschluss über den vorrangig zu verbessernden Bereich. Wenn Sie vereinfacht dargestellt werden, sind die Resultate der Wandelbarkeitsanalyse wahrscheinlich gut gegenüber Menschen zu kommunizieren, welche sich nicht hauptberuflich mit Thermodynamik beschäftigen, da die zugrunde liegenden Größen sich an die allgemein bekannten Konzepte von Energie und Qualität anlehnen lassen.

Ein Nachteil des Wandelbarkeitskonzepts und der Wandelbarkeitsmethode liegt in der aufwändigeren Berechnung und Bilanzierung begründet. Dies begrenzt wahrscheinlich die Anwendung des Konzepts bis dezidierte Software-Werkzeuge zur Anwendung von Wandlungskonzept und -methode zur Verfügung stehen. Zusätzlich, lässt die über das Exergiekonzept hinausgehende Komplexität der Theorie es zweifelhaft erscheinen, dass das Konzept auf großer Breite tiefergehend verstanden wird. Dennoch kann die Beschäftigung mit dieser Theorie helfen, die Größe Exergie besser zu verstehen und damit besser anzuwenden.

Zusammenfassend kann festgestellt werden, dass Wandelbarkeitsmethode und -konzept das Potential zu haben scheinen, ein nützlicher Beitrag zur thermodynamischen Theorie zu werden. Sie ermöglichen eine neue, wissenschaftlich fundierte und verständliche Perspektive auf die Exergie, bilden die Grundlage für innovative Darstellungsformung von Analyseergebnissen und können Ausgangspunkt für eine verbesserte Kommunikation des Exergiekonzepts und seiner Abgrenzung zum Energiekonzept werden.

Nomenclature

Lower case letters of upper case variables denote mass specific units unless defined otherwise.

c	velocity	[m/s]
COP	coefficient of performance	$[-]$
En	energy	$[J]$
E	exergy	$[J]$
f	factor	$[-]$
H	enthalpy	$[J]$
HHV	higher heating value	$[J]$
h	specific enthalpy	[J/kg]
g	gravitational acceleration	[m/s²]
m	mass	$[kg]$
M	molar mass	[kg/mol]
n	number of moles	$[mol]$
p	pressure	$[Pa]$
Q	heat	$[J]$
W	work	$[J]$
x	mole fraction in general	$[-]$
y	ratio	$[-]$
z	altitude	$[m]$

Greek letters

Δ	[Delta] difference between input and exit
η	[eta] energy efficiency
ε	[epsilon] exergetic efficiency
ν	[nu] stoichiometric coefficient
Σ	[Sigma] total, sum over all elements
σ	[sigma] entropy generation, exergetic sensitivity
τ	[tau] transformability, as a subscript transformation
ξ	[xi] transformability efficiency

Nomenclature

Superscripts

o	property at absolute values (reference temperature is 0 K)
$*$	compensation, alternative
$-$	bar over symbol denotes property on a molar basis
$\cdot$	dot over symbol denotes time rate
0	property at standard state for measurements of formation properties
CH	chemical
EL	electrical
id	ideal
KN	kinetic
M	mechanical
mx	maximal
mn	minimal
N	nonreactive
PH	physical
PT	potential
H	effective thermal, associated with an energy transfer due to enthalpy differences of a mass flow
Q	associated with a conductive heat transfer
R	reactive
T	thermal
TO	total
X	a specific type of (such as mechanical, thermal ...)

Subscripts

0	at the condition of the exergy reference environment. For conductive heat flows and transformation energy flows associated with such heat flows: heat flows at reference temperature
a	average
aF	attributed combustible fuel
cF	combustible fuel (only used if subscript F is used for total fuel in efficiency-like ratios)

Nomenclature

CHP	combined heat and power
D	destruction
DH	district heating
dr	driving
E	exergy
En	energy
e	exit
ef	energy flow
f	formation
F	fuel
gr	ground
gt	geothermal
h	high temperature
HC	high temperature cascade
i	input
j	variable, substance indicator, component indicator
l	low temperature
L	loss
LC	low temperature cascade
LE	LowEx
mf	mass flow
NG	natural gas
P	product
p	pressure
R	reactand
r	room
RF	reference technology
rq	required
rv	internally reversible
$T0$	reference temperature but not reference pressure
tr	transport, transfer
U	useful
wt	waste

Nomenclature

Abbreviations

CHP	combined heat and power
DH	district heating
HHV	higher heating value
IEA	International Energy Agency

List of Figures

2.1	Energy flow chart of a reversible heat engine process	26
2.2	Exergy and transformation energy flow charts of a reversible power cycle operating between reference temperature and a temperature below reference temperature	29
2.3	Energy and mass flow charts of a reversible power cycle used for transformation energy derivation for mass flows at temperatures below reference temperature	31
2.4	Energy and mass flow chart of a reversible heat engine process	33
2.5	Flow charts of a reversible fuel cell process used for reactive transformability derivation	37
2.6	Flow charts of a reversible fuel cell process used for the derivation of nonreactive transformation energy	40
2.7	Exergy flow chart of a reversible fuel cell for substances above reference condition	42
3.1	Flow charts of a reversible heat engine process operating between reference temperature and a mass flow at a temperature below reference temperature - part 1	49
3.2	Flow charts of a reversible heat engine process operating between reference temperature and a mass flow at a temperature below reference temperature - part 2	49
4.1	Transformation energy efficiency - Transformability ratio diagram for heat supply systems at different reference temperatures	75
4.2	Transformation energy efficiency - Transformability ratio diagram for cooling systems at different reference temperatures	77
4.3	Vapor-compression cascade refrigeration machine	85
4.4	Reference scenario of the ExergyFingerprint of an old building supplied by the average german power mix and heat from a gas condensing boiler	89
4.5	ExergyFingerprint of an old building supplied by the average German power mix and heat from a block heat and power plant.	90
4.6	ExergyFingerprint of an insulated building supplied by the average German power mix and heat from a gas condensing boiler.	91
A.1	Flow charts of a heat supply system using a heat pump	127
A.2	Flow chart of the separation of a CHP process into subsystems	137
A.3	Flow charts of a heat supply system based on a gas condensing boiler	145
A.4	Flow charts of a heat supply system based on a geothermal source	147
A.5	Flow charts of a heat supply system based on a heat pump	148
A.6	Flow charts of a heat supply system based on a block heat and power plant	151
A.7	Flow charts of a cold supply system based on a compression refrigeration machine - part 1	154
A.8	Flow charts of a cold supply system based on a compression refrigeration machine - part 2	155
A.9	Flow charts of a cold supply system based on seawater cooling - part 1	157
A.10	Flow charts of a cold supply system based on direct seawater cooling - part 2	158

List of Figures

A.11 Flow charts of a cold supply system based on an absorption refrigeration machine - part 1 159

A.12 Flow charts of a cold supply system based on an absorption refrigeration machine - part 2 159

A.13 Model of an absorption refrigeration machine as a combination of a heat engine and a compression refrigeration machine 161

List of Tables

4.1	Results of the evaluation of exemplary heating systems	68
4.2	Results of the evaluation of exemplary cooling systems	71
4.3	Results of the evaluation of exemplary heating systems at different reference states	74
4.4	Results of the evaluation of exemplary cooling systems at different reference states	76
4.5	Results of the evaluation of basic processes	80
4.6	Results of the transformability analysis of a vapor-compression cascade refrigeration machine	86
4.7	Effective thermal transformabilities, average transformabilities and effective compensation heat flows in the vapor-compression cascade refrigeration machine	87
A.1	Thermodynamic data of air for the assessment of an exemplary heat pump	127
A.2	Overview on transformation energy and transformability associated with various types of mass and energy flows - part 1	130
A.3	Overview on transformation energy and transformability associated with various types of mass and energy flows - part 2	131
A.4	General expressions for effective thermal transformabilities	132
A.5	General assumptions for the comparative evaluation of heat supply systems	144
A.6	Specific assumptions for the evaluation of a heat supply system based on a condensing boiler	145
A.7	Specific assumptions for the evaluation of a heat supply system based on geothermal district heating	146
A.8	Specific assumptions for the evaluation of a heat supply system based on an electrical ground-source heat pump	148
A.9	Specific assumptions for the evaluation of a heat supply system based on a block heat and power plant	150
A.10	General assumptions for the comparative evaluation of cooling systems	153
A.11	Specific assumptions for the evaluation of a cooling supply system based on a compression refrigeration machine	154
A.12	Specific assumptions for the evaluation of a seawater cooling system	157
A.13	Specific assumptions for the evaluation of a cooling supply system based on an absorption refrigeration machine operated with waste heat	160
A.14	Exemplary daily exergy values for a conductive heat flow of $1\,kW$ at a temperature of $340\,K$	163
A.15	Assumptions for the analysis of some basic processes	165
A.16	Expressions for average transformabilities of heat exchangers, boilers, heat pumps and refrigeration machines	166
A.17	Expressions for average transformabilities of heat engines, expanders and compressors	167
A.18	Equations for the analysis of heat exchangers, boilers, heat pumps and refrigeration machines	168

List of Tables

A.19 Equations for the analysis of heat engines, expanders and compressors 169
A.20 Thermodynamic data of air 170
A.21 Thermodynamic data of ethane (R170), the working fluid of the low cascade 170
A.22 Thermodynamic data of propane (R290), the working fluid of the high cascade . . . 170
A.23 Exergy, transformation energy and compensation heat flows associated with air flows in the vapor-cascade refrigeration machine 171
A.24 Exergy, transformation energy and compensation heat flows associated with ethane flows in the vapor-cascade refrigeration machine 171
A.25 Exergy, transformation energy and compensation heat flows associated with propane flows in the vapor-cascade refrigeration machine 171
A.26 Ideally required heat flows 172
A.30 Basic data for the calculation of the exemplary ExergyFingerprint demand structure . 173
A.27 Effective thermal transformabilities and effective compensation heat flows - equations 174
A.28 Average in- and output transformabilities - equations 175
A.29 Equations for the evaluation of a vapor-compression cascade refrigeration machine . 176
A.31 Required temperature levels and transformabilities as basic data for the exemplary ExergyFingerprints 177

Bibliography

AHRAE: 1997, *1997 ASHRAE Handbook: Fundamentals*, Vol. SI Edition. American Society of Heating, Refrigerating and Air-Conditioning Engineers, Inc.

Ahrendts, J.: 1977, *Die Exergie chemisch reaktionsfähiger Systeme (in German)*, No. 579 in VDI-Forschungsheft. VDI-Verlag, Düsseldorf.

Ahrendts, J.: 1980, 'Reference states'. *Energy* (5), 667–677.

Balocco, C., S.Papeschi, G.Grazzini, and R. Basosi: 2003, 'Using exergy to analyze the sustainability of an urban area'. *Ecological Economics* (48), 231 – 244.

Bejan, A., G. Tsatsaronis, and M. J. Moran: 1996, *Thermal Design and Optimization*. John Wiley and Sons, Inc.

Berthiaume, R., C. Bouchard, and M. A. Rosen: 2001, 'Exergetic evaluation of the renewability of a biofuel'. *Exergy an International Journal* 1(4), 256–268.

Bittrich, P. and D. Hebecker: 1999, 'Classification and Evaluation of heat transformation processes'. *International Journal of Thermal Science* (38), 465–474.

BMWI: 2008, 'Energiedaten 2006 / Tabelle 7a (in German)'. *Bundesministerium für Wirtschaft und Technologie* www.bmwi.de/BMWi/Redaktion/Binaer/Energiedaten/ energiegewinnung-und-energieverbrauch5-eev-nach-anwendungsbereichen,property=blob, bereich=bmwi,sprache=de,rwb=true.xls (Accessed on September 06th, 2008).

Bosnjakovic, F. and K. Knoche: 1998, *Technische Thermodynamik: Teil 1 (in German)*. Steinkopff Verlag, 8th edition.

Bosnjakovitch, F.: 1935, *Technische Thermodynamik Bd. 1 (in German)*.

Cengel, Y. A. and M. A. Boles: 2006, *Thermodynamics: an engineering approach*. McGraw-Hill, 6th edition.

Chang, H. and S.-C. Chuang: 2003, 'Process analysis using the concept of intrinsic and extrinsic exergy losses'. *Energy* (28), 1203–1228.

DeStatis: 2009, 'Bevölkerungsstand (in German)'. *Statistisches Bundesamt Deutschland* www.destatis.de/jetspeed/portal/cms/Sites/destatis/Internet/DE/Navigation/ Statistiken/Bevoelkerung/Bevoelkerungsstand/Bevoelkerungsstand.psml (Accessed on April 14th, 2009).

Dincer, I. and Y. A. Cengel: 2001, 'Energy, Entropy and Exergy Concepts and Their Roles in Thermal Engineering'. *Entropy* (2), 116–149.

Dittman, A., T. Sander, and G. Menzler: 2009, 'Die ökologische Bewertung von Wärme und Elektroenergie ein Instrument zur Erhöhung der Akzeptanz der Kraft-Wärme-Kopplung (in German)'. *VIK Mitteilungen - Sonderdruck*.

Dunning-Davies, J.: 1965, 'Carathéodory's Principle and the Kelvin Statement of the Second Law'. *Nature* (208), 576–577.

Bibliography

enob.info: 2009, 'LowEx: Heizen und Kühlen mit Niedrig-Exergie (in German)'. *Forschung für Energieoptimiertes Bauen* **www.enob.info/de/forschungsfelder/lowex/** (Accessed on July 6th, 2009).

Erlach, B., G. Tsatsaronis, and F. Cziesla: 2001, 'A new approach for assigning costs and fuels to cogeneration products'. *International Journal of Applied Thermodynamics* **4**(3), 145 – 156.

Franke, U.: 1998, 'Prozessbewertung ohne Exergie (in German)'. Technical report, FH Flensburg.

Fratzscher, W.: 1997, 'Exergy and possible applications'. *Revue Generale de Thermique* (36), 690–696.

Fratzscher, W., V. M. Brodjanskij, and K. Michalek: 1986, *Exergie (in German)*. VEB Deutscher Verlag für Grundstoffindustrie.

Grassmann, P.: 1951, 'Technische Arbeitsfähigkeit als praktische Rechengröße (in German)'. *Zeitschrift für Wärme- und Kältetechnik* **2**(8), 161–166.

Hau, J. L.: 2005, 'Toward environmentally conscious process systems engineering via joint thermodynamic accounting of industrial and ecological systems'. Ph.D. thesis, The Ohio State University.

Hebecker, D., P. Bittrich, and K. Riedl: 2004, 'Hierarchically Structured exergetic and exergoeconomic analysis and evaluation of energy conversion processes'. *Energy Conversion and Management* (46), 1247 – 1266.

Ignatenko, O., A. van Schaik, and M. Reuter: 2007, 'Exergy as a tool for evaluation of the resource efficiency of recycling systems'. *Minerals Engineering* (20).

Jentsch, A., C. Dötsch, C. Beier, and S. Bargel: 2009, 'ExergyFingerprint - Neues Bewertungswerkzeug für Energieversorgungsszenarien (in German)'. *EuroHeat&Power (German)* (04).

Jorgensen, S. E.: 1999, 'State-of-the-art ecological modelling with emphasis on development of structural dynamic models'. *Ecological Modelling* (120), 75–96.

Kelly, S., G. Tsatsaronis, and T. Morosuk: 2009, 'Advanced exergetic analysis: Approaches for splitting the exergy destruction into endogenous and exogenous parts'. *Energy* (34).

Klenner, S.: 2008, 'Jahresarbeitszahlen von 33 Wärmepumpen (In German)'. *Erdwärme-Zeitung Online* **www.erdwaerme-zeitung.de/meldungen/jahresarbeitszahlenvon33waermepumpen10023.html** (Accessed on September 27th, 2008).

Lazzareto, A. and G. Tsatsaronis: 2006, 'SPECO: A systematic and general methodology for calculating efficiencies and costs in thermal systems'. *Energy* (31), 1257–1298.

Lems, S., H. van der Kooi, and J. de Swan Aarons: 2003, 'Quantifying technological aspects of process sustainability'. *Clean Technology and Environmental Policy* (5), 248–253.

lowex.info: 2009, 'Verbundvorhaben LowEx (in German)'. **www.lowex.info/lowex.html** (Accessed on July 6th, 2009).

Bibliography

LowEx.net: 2009, 'Our Mission'. www.lowex.de (Accessed on July 6th, 2009).

Lukas, K.: 2004, *Thermodynamik (in German)*. Springer Verlag, 4th edition.

Machat, M. and K. Werner: 2007, 'Entwicklung der spezifischen Kohlendioxid-Emissionen des deutschen Strommix (in German)'. *Climate Change (1)* www.umweltdaten.de/publikationen/fpdf-l/3195.pdf.

Meyer, L., G. Tsatsaronis, J. Buchgeister, and L. Schebek: 2009, 'Exergoenvironmental analysis for evaluation of the environmental impact of energy conversion systems'. *Energy* (34), 75–89.

Moran, M. J. and H. N. Shapiro: 2000, *Fundamentals of Engineering Thermodynamics*. Von Hoffmann Press, 4th edition.

Moran, M. J. and H. N. Shapiro: 2007, *Fundamentals of Engineering Thermodynamics*. John Wiley and Sons, Inc., 6th edition.

Morosuk, T. and G. Tsatsaronis: 2008, 'A new approach to the exergy analysis of absorption refrigeration machines'. *Energy* (33), 890–907.

Nesheim, S. J. and I. S. Ertesvag: 2007, 'Efficiencies and indicators defined to promote combined heat and power'. *Energy Conversion and Management* (48), 1004–1015.

Nieuwlaar, E. and D. Dijk: 1993, 'Exergy Evaluation of Space Heating Options'. *Energy* **18**(7), 779–790.

NIST: 2007, 'Refprop 8.0 - Reference Fluid Thermodynamic and Transport Properties'. *National Institute of Standards and Technology*.

Petela, R.: 2003, 'Exergy of undiluted ratiation'. *Solar Energy 74* (74), 469 – 488.

Phylipsen, G., K. Blok, and E. Worell: 1998, 'Handbook on international comparisons of energy efficiency in the manufacturing industry'. Ph.D. thesis, Utrecht University.

Radgen, P. and J. Oberschmidt: 2006, 'Multidimensional assessment of heat and power supply technologies with a special focus on CHP'. *Fraunhofer ePrints*.

Rant, Z.: 1956, 'Exergie - ein neues Wort für technische Arbeitsfähigkeit (in German)'. *Forschung auf dem Gebiete des Ingenieurswesens* **22**(1), 36–37.

Riedl, K.: 2006, 'Exergetische und Exergoökonomische Bewertungsverfahren von Verfahren der Energie- und Stoffwandlung (in German)'. Ph.D. thesis, Martin-Luther Universität Halle-Wittenberg.

Rosen, M. A.: 2002, 'Assessing energy technologies and environmental impacts with the principles of thermodynamics'. *Applied Energy* (72), 427–441.

Rosen, M. A.: 2008a, 'Allocating carbon dioxide emissions from cogeneration systems: descriptions of selected output-based methods'. *Journal of Cleaner Production* (16), 171–177.

Rosen, M. A.: 2008b, 'Role of exergy in increasing efficiency and sustainability and reducing environmental impact'. *Energy Policy* (36), 128–137.

Bibliography

Rosen, M. A. and I. Dincer: 2004, 'Effect of varying dead-state properties on energy and exergy analyses of thermal systems'. *International Journal of Thermal Sciences* (43), 121–133.

Stephan, K. and F. Mayinger: 1986, *Thermodynamik: Band 1 Einstoffsysteme (in German)*. Springer Verlag, 12th edition.

Susani, L., F. Pulselli, S. E. Jorgensen, and S. Basianoni: 2005, 'Comparison between technological and ecological exergy'. *Ecological Modelling* (193), 447 – 456.

Szargut, J.: 2005, *Exergy Method - Technical and Ecological Applications*. WITPress Southampton, Boston.

Szargut, J., D. R. Morris, and F. R. Steward: 1988, *Exergy Analysis of Thermal, Chemical, and Metallurgical Processes*. Hemisphere Publishing / Springer-Verlag.

Tsatsaronis, G.: 1984, 'Combination of Exergetic and Economic Analysis in Energy-Conversion Processes'. *Energy Economics and Management in Industry* **Proceedings of the European Congress, Algarve, Portugal, April 2-5, Vol. 1,** 151–157.

Tsatsaronis, G.: 1999, 'Strengths and Limitations of Exergy Analysis'. *Thermodynamic Optimization of Complex Energy Systems* pp. 93–100.

Tsatsaronis, G.: 2007, 'Definitions and nomenclature in exergy analysis and exergoeconomics'. *Energy* (32), 249–253.

Tsatsaronis, G., J. Bausa, C. F., J.-B. Eggers, K. K., and K. A.: 2007, *Umdruck zur Vorlesung: Energietechnik (in German)*. TU Berlin.

Tsatsaronis, G. and M.-H. Park: 2002, 'On avoidable and unavoidable exergy destructions and investment costs in thermal systems'. *Energy Conversion and Management* (43), 1259–1270.

Tsatsaronis, G. and M. Winhold: 1985, 'Exergoeconomic Analysis and Evaluation of Energy Conversion Plants. Part I-A New General Methodology'. *Energy-The International Journal* (10), 69–80.

Utlu, Z. and A. Hepbasli: 2007, 'Parametrical investigation of the effect of dead (reference) state on energy and exergy utilization'. *Renewable and Sustainable Energy Reviews* **11**(4), 603–634.

VTT: 2003, *Low exergy systems for heating and cooling of buildings - Heating and cooling with focus on increased energy efficiency and improved comfort*. VTT Technical Research Centre of Finland.

Wall, G.: 1977, 'Exergy - A useful concept within ressource accounting'. Technical report, Chalmers University of Technology and University of Göteborg.

Wulf, J. D., H. van Langenhove, J. Mulder, M. van den Berg, H. van der Kooi, and J. de Swaan Aarons: 2000, 'Illustrations towards quantifying the sustainability of technology'. *Green Chemistry* (2), 108–114.

Xia, C., Y. Zhu, and B. Lin: 2008, 'Renewable energy utilization evaluation method in green buildings'. *Renewable Energy* (33), 883–886.

Appendices

A1 Summary of the exergy interpretation underlying this work

To simplify the understanding of the interpretation of exergy underlying this work, a short outline of exergy basics and common misconceptions is given below. It is essentially following the interpretation of exergy provided in Bosnjakovic and Knoche (1998):

1. Only exergy associated with mass flows and energy transfers is considered in this work. Exergy associated with closed systems is not discussed. Therefore, the term exergy used in this work refers only to exergy associated with transfers.

2. Exergy is the maximum work potential associated with a combined system of a flow of interest that is assumed to be provided at constant properties and a defined thermodynamic environment.

3. Assuming a constant reference environment an exergy rate can be associated with any energy or entropy transfer that has different intensive parameters than the environment. The association is a purely practical measure since exergy is always a function of a combined system. The association of exergy with the flow does not indicate that if work were to be generated from the combined system that its source is the transfer under consideration.

4. Negative values of exergy flows are interpreted as being opposed to the transfer they are associated to. This allows a consistent consideration of such flows in the exergy balance.

5. If exergy is associated with an enthalpy transfer[3] or a mass-free energy flow it cannot exceed the value of the enthalpy flow or the energy transfer it is associated with. If its absolute value is larger it has to be prefaced with a negative sign.

6. Exergy evaluation and analysis should always be based on the separate evaluation of the different types of exergy (thermal, mechanical, chemical...), since the use of total exergy or physical exergy can result in difficulties if using exergetic efficiency for evaluation - see appendix A3 on page 126 ff. for a discussion.

A1.1 Avoiding misconceptions

The misconceptions presented here can be found in various sources concerning exergy. However, some of those misconception are shared by so many people that a complete enumeration of those using them does not appear to be sensible. Therefore no references are given here.

1. Concerning „Exergy of ..."
    - Exergy is not a property of energy transfers or mass flows. It is a property of the combined system of energy or mass transfer of interest and the environment. Exergy can only be associated with an energy or mass transfer, if the relevant specifications of the environment

---

[3] It is assumed that the reference state for the calculation of enthalpy is assumed to be equal to the reference state assumed for the calculation of exergy.

Appendices

are known. Therefore, the commonly used expression „exergy of" is replaced with „exergy associated with".

2. Concerning the interpretation of exergy as useful energy.

- Exergy is not a form of energy, but a *potential* to generate work from the combined system of flow and environment. Compressed gas flows at reference temperature are associated with exergy although they only differ significantly from the environment in terms of specific entropy while not having a significantly different specific enthalpy. As a consequence the notion of exergy as a „share of energy" contained in the energy flow is not followed.

3. Concerning the use of anergy.

- Anergy is a concept that is problematic and not fully consistent if applied to heat or mass transfers at temperatures below reference temperature and to mechanical and nonreactive exergy (Bosnjakovic and Knoche, 1998; Szargut, 2005). It will therefore not be used.

4. Concerning interpretation of exergy as a measure of energy quality.

- Exergy is not a measure of energy quality of an energy or mass flow under consideration as it is always influenced by quantitative effects. Exergy is considered a combined measure of thermodynamic quality and quantity.

5. Concerning the labelling of exergy analysis as second law analysis.

- Exergy analysis does not equal second law analysis since exergy analysis always incorporates aspects of the first law of thermodynamics as well. Therefore, the results of exergy analysis are always influenced by external losses of energy and matter. Consequently, the synonymous use of second-law efficiency and exergetic efficiency is avoided.

A2 Calculation of chemical exergy

The calculation of chemical exergy $\dot{E}_j^{CH}$ for substances not present in the environment is based on the specific molar enthalpy of formation $\bar{h}_f$, the specific molar absolute entropy $\bar{s}^o$ and the specific molar enthalpy $\bar{h}$ of a considered molecule j at a certain temperature T, pressure p and mole fraction x. Enthalpy of formation is a thermochemical property and according to Moran and Shapiro (2007) defined as a function the enthalpy of formation at the standard state $\bar{h}_{f_j}^0$, which is usually defined as $298\,K$ and $1\,atm$ and the specific molar enthalpy $\bar{h}_j$ at the standard state and at the considered state:

$$\bar{h}_{f_j}(T_0,p_0,x_j) = \bar{h}_{f_j}\left(T^0,p^0,x_j^0\right) + [\bar{h}_j(T_0,p_0,x_j) - \bar{h}_j\left(T^0,p^0,x_j^0\right)] = \bar{h}_{f_j}^0 + \Delta\bar{h}_j \quad \text{(a.1)}$$

The absolute entropy is defined as (Moran and Shapiro, 2007) as a function of the specific molar absolute entropy at standard state and the specific molar entropy $\bar{s}_j$ of the substance:

$$\bar{s}^o{}_j(T_0,p_0,x_j) = \bar{s}^o{}_j\left(T^0,p^0,x_j^0\right) + [\bar{s}_j(T_0,p_0,x_j) - \bar{s}_j\left(T^0,p^0,x_j^0\right)]$$

Appendices

A general hydrocarbon (indicated by C_aH_b) - oxygen reaction is characterized by the following reaction equation:

$$C_aH_b + \left(a + \frac{b}{4}\right)O_2 \rightarrow aCO_2 + \frac{b}{2}H_2O$$

The chemical exergy flow associated with a hydrocarbon flow can therefore be expressed as (Moran and Shapiro, 2007):

$$\begin{aligned}
\dot{E}^{CH}_{C_aH_b}(T_0, p_0) &= \dot{n} \cdot \left[\bar{h}_{fC_aH_b} + \left(a + \frac{b}{4}\right)\bar{h}_{fO_2} - a\bar{h}_{fCO_2} - \frac{b}{2}\bar{h}_{fH_2O}\right](T_0, p_0) \\
&- \dot{n} \cdot \left\{T_0 \cdot \left[\bar{s}^o_{C_aH_b} + \left(a + \frac{b}{4}\right)\bar{s}^o_{O_2} - a\bar{s}^o_{CO_2} - \frac{b}{2}\bar{s}^o_{H_2O}\right](T_0, p_0)\right\} \\
&= H\dot{H}V_{C_aH_b}(T_0, p_0) \\
&- \dot{n} \cdot \left\{T_0 \cdot \left[\bar{s}^o_{C_aH_b} + \left(a + \frac{b}{4}\right)\bar{s}^o_{O_2} - a\bar{s}^o_{CO_2} - \frac{b}{2}\bar{s}^o_{H_2O}\right](T_0, p_0)\right\} \\
&+ a\bar{e}^N_{CO_2} - \left(\frac{b}{2}\right)\cdot\bar{e}^N_{H_2O_{(l)}} - \left(a + \frac{b}{4}\right)\cdot\bar{e}^N_{O_2}
\end{aligned}$$

Using the stoichiometric factor ν_j and extending the equation to all combustible fuels (index cF) instead of C_aH_b a more general equation of the specific molar chemical exergy $\bar{e}^{CH}_F$ can be obtained. Since the stoichiometric factor of the combustible fuel within the reaction equation is not always 1 but a reaction specific property it appears sensible to use the ratio $\frac{\nu_j}{\nu_F}$ instead of ν_j as a factor for reactands and products. Using this factor the molar specific chemical exergy of a fuel can be expressed as:

$$\bar{e}^{CH}_F(T_0, p_0) = \left(\bar{h}_{fF} - T_0\bar{s}^o_F\right) + \sum_R \frac{\nu_R}{\nu_F}\cdot\left(\bar{h}_{fR} - T_0\bar{s}^o_R\right) - \sum_P \frac{\nu_P}{\nu_F}\cdot\left(\bar{h}_{fP} - T_0\bar{s}^o_P\right) \quad \text{(a.2)}$$

If the considered substance is present within the environment chemical exergy which then equals nonreactive chemical exergy can be calculated on a mass basis. The nonreactive exergy associated with a fluid can be obtained from the concentration difference of the fluid as present in the considered mixture with mole fraction x_j to the concentration of the substance in the environment with mole fraction $x_{j,0}$. The nonreactive exergy $\dot{E}^N_j$ can be defined as:

$$\dot{E}^N_j = \dot{m}_j \cdot \{[h_j(T_0, p_0, x_j) - h_j(T_0, p_0, x_{j,0})] - T_0 \cdot (s_j(T_0, p_0, x_j) - s_j(T_0, p_0, x_{j,0}))\} \quad \text{(a.3)}$$

Normally, nonreactive exergy cannot be used to generate work in thermal processes (Bejan et al., 1996).

The chemical exergy of real mixtures is usually cumbersome to calculate. Therefore, for some mixtures, e.g. lignite, numerical equations exist that allow the approximation of the reactive exergy associated with a fuel based on the mass contributions of the constituents. (Moran and Shapiro, 2007).

To simplify the calculation of chemical exergy the concept of standard chemical exergy can be used (Fratzscher et al., 1986; Bejan et al., 1996; Szargut, 2005; Moran and Shapiro, 2007). The standard

Appendices

chemical exergy gives values for chemical exergy at $T_0^0 = 298,15$ and $p_0^0 = 1\,atm$ in a chemical environment that is close to the composition of the natural environment. The advantage of the standard chemical exergy approach is that it allows the calculation of values for chemical exergy-based on tabulated data and entropy and enthalpy differences between the standard state and the considered state. The effects of slight variations in temperature and pressure from the standard reference state on chemical exergy can usually be neglected so that for many engineering applications it is sufficient to use the standard chemical exergy values (Bejan et al., 1996). For further details on the application and for values of standard chemical exergy see the provided references.

Finally, it has to be mentioned that the reference state plays a great role for the results of the assessment of chemical exergy since some substances, such as sulfur, are not present in significant concentrations in the environment. Different reference environment models can be used to assess chemical exergy. e.g. (Bejan et al., 1996; Moran and Shapiro, 2000) use two models: model I by Ahrendts (1980) with $p_0 = 1,019\,atm$ and model II by Szargut et al. (1988) with $p_0 = 1\,atm$. Model I imposes that the reference environment is in mutual equilibrium and in restricted equilibrium for nitric acid and nitrates. The chemical composition of the gas phase models the real composition of the natural atmosphere. Model II defines the reference environment based on reference substances that are abundantly present in the environment. This approach leads to an environment which is not for all components in total thermodynamic equilibrium. However, this reference environment comes closer to real composition of the natural environment than model I but has the disadvantage that due to its deviations from thermodynamic equilibrium work can in principle be generated from the environment itself. Since both models provide significantly different values for chemical exergy, it is mandatory to only use one definition of reference environment when comparing chemical exergy values.

A3 On the use of separate types of exergy for exergy analysis

Lazzareto and Tsatsaronis (2006) indicate that a separate assessment of the different exergy types would lead to greater accuracy in exergetic analysis. To illustrate the problem of the use of physical exergy for the analysis of components, an exemplary exergy analysis is performed on a heat pump with heat source and heat discharge from or to air. The following example has been chosen, because it allows the discussion of some aspects of the theoretical basics of the exergetic efficiency definition. Operation parameters have been chosen to illustrate the necessity of a separate evaluation and do not represent common parameters of real heat pumps.

The flow charts of the heat pump are shown in Figure A.1. The data on which the analysis is based is summarized in Table A.1. The reference state is defined by $p_0 = 0,1013\,MPa$ and $T_0 = 298,15\,K$.

Appendices

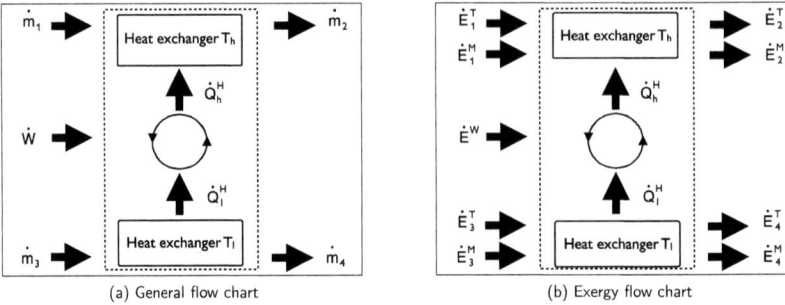

(a) General flow chart (b) Exergy flow chart

Figure A.1: Flow charts of a heat supply system using a heat pump

Table A.1: Thermodynamic data of air for the assessment of an exemplary heat pump

Flow	$\dot{m}$	T	p	h	h_{T0}	s	s_{T0}	e^{PH}	e^T	e^M
	$\dfrac{kg}{s}$	K	MPa	$\dfrac{kJ}{kg}$	$\dfrac{kJ}{kg}$	$\dfrac{kJ}{kg \cdot K}$	$\dfrac{kJ}{kg \cdot K}$	$\dfrac{kJ}{kg}$	$\dfrac{kJ}{kg}$	$\dfrac{kJ}{kg}$
1	1,0	353,2	0,4052	353,4	297,8	6,6	6,5	123,17	4,58	118,59
2	1,0	383,2	0,2026	384,1	298,2	6,9	6,7	69,64	10,33	59,31
3	0,5	353,2	0,4052	353,4	297,8	6,6	6,5	123,17	4,58	118,59
4	0,5	303,2	0,2026	303,3	298,2	6,7	6,7	59,35	0,04	59,31

The heat pump is considered to be thermally ideal while large pressure drops occur in the heat exchangers. The work flow input $\dot{W}$ required by the heat pump can be obtained from the energy balance of the heat pump as a function of the mass flows $\dot{m}$ and the specific enthalpies h as:

$$\dot{W} = \dot{m}_1 \cdot (h_2 - h_1) + \dot{m}_3 \cdot (h_3 - h_4)$$
$$= 3,371 \, kW$$

Based on the data in Table A.1 and the provided work it should in principle be possible to calculate exergetic efficiency using the physical exergy values. Exergetic efficiency is defined as the ratio of the exergy flow associated with the useful product to the exergy flow associated with the fuel exergy. Fuel and product exergy are defined according to the following rules by Lazzareto and Tsatsaronis (2006) as follows [4]:

The product is defined to be equal to the sum of

---
[4] Deviating from the definition given, the considered definition does not allow to subtract all exergy increases (between inlet and outlet) that are not in accord with the purpose of the component from the fuel term. This modification is necessary to avoid nonsensible results in the transformability analysis, such as average transformabilities above 100 %.

Appendices

- all the useful exergy flows to be considered at the outlet (including the exergy of energy streams generated in the component) plus
- all the exergy flow increases between inlet and outlet (i.e. the exergy additions to the respective material streams) that are in accord with the purpose of the component.

Similarly, the fuel is defined to be equal to

- all the exergy flows to be considered at the inlet (including the exergy of energy streams supplied to the component) plus
- all the exergy flow decreases between inlet and outlet (i.e. the exergy removals from the respective material streams)

If considering Table A.1 it becomes apparent that mechanical exergy associated with flow 1 decreases while the the thermal exergy associated with the same flow increases. Since the mechanical exergy decrease is significantly larger than the thermal exergy increase the physical exergy associated with flow 1 is decreased by the process. Thus, the change of physical exergy associated with mass flow 1 could not be defined the product of the process. Since physical exergy also decreases in flow 3 and the work flow is an input it is impossible to define exergetic efficiency for the considered heat pump in accord with the rules for the definition of fuel and product in exergetic efficiency on the basis of consideration of physical exergy.

Alternatively, for the same application the exergetic efficiency ε can be defined taking thermal $\dot{E}^T$ and mechanical exergy flows $\dot{E}^M$ into account separately. Since the thermal exergy flow associated with mass flow 1 increases, it can be considered the product of the heat pump. The decreases in mechanical exergy and the work flow constitute the fuel term. Consequently this efficiency could be defined in accordance with Lazzareto and Tsatsaronis (2006) as:

$$\begin{aligned}\varepsilon &= \frac{\dot{E}_2^T - \dot{E}_1^T}{\dot{E}_3^T - \dot{E}_4^T + \dot{E}_1^M - \dot{E}_3^M + \dot{E}_2^M - \dot{E}_1^M + \dot{W}}\\ &= 5,9\,\%\end{aligned}$$

The example demonstrates that if one type of exergy increases between input and output while another one decreases (as for flow 1 - 2) then the use of the physical exergy difference can make it impossible to sensibly define exergetic efficiency. As a consequence, it can be concluded that a separate evaluation of mechanical and thermal exergy is necessary if operating with exergy differences in general, since it is necessary to evaluate first whether the different types of exergy increase or decrease from input to output before a sensible definition of the product and consequently exergetic efficiency can be found.

In subchapter 3.2.2 ff. the necessity of limiting fuel to the sum of exergy inputs and exergy increases has been explained with the need to ensure sensible definitions of the average input transformability. Including this aspect allows a generalization of the results obtained from the consideration of the presented example.

A separate evaluation of all exergy types allows to decide for every exergy type separately whether the value is increasing or decreasing within the process, thus avoiding the summation of increasing and decreasing exergy types in fuel or product definition. In such a way eventual problems with the universality of the assessment and nonsensible definitions of average input transformability can be avoided with certainty at the cost of a more complicated calculation scheme. A separate consideration of the different types of exergy is therefore required to guarantee a universal and consistent evaluation using transformability analysis and evaluation method.

A4 Summary of transformability and transformation energy definitions for practical application

For quick reference the following tables summarizes the various transformabilities and transformation energies that become relevant if the mass or energy flow under consideration has intensive properties different from those of the environment.

Table A.2: Overview on transformation energy and transformability associated with various types of mass and energy flows - part 1

Type of flow	Property of transfer which is different from the environment	Transformability τ_i	Transformation energy flow $\dot{E}n_\tau$	compensation heat flow $\dot{Q}_0^*$
Conductive heat flow (Thermal exergy)	$T \geq T_0$	$1 - \dfrac{T_0}{T}$	$\dot{Q}$	0
Conductive heat flow (Thermal exergy)	$T < T_0$	$1 - \dfrac{T}{T_0}$	$-\dfrac{T_0}{T} \cdot \dot{Q}$	$\left(1 + \dfrac{T_0}{T}\right) \cdot \dot{Q}$
Mass flow (Thermal exergy)	$T \geq T_0$ at $p = const.$	$1 - \dfrac{T_0 \cdot (s - s_{T0})}{(h - h_{T0})}$	$\dot{m} \cdot (h - h_{T0})$	0
Mass flow (Thermal exergy)	$T < T_0$ at $p = const.$	$1 - \dfrac{(h - h_{T0})}{T_0 \cdot (s - s_{T0})}$	$-\dot{m} \cdot T_0 \cdot (s - s_{T0})$	$\dot{m} \cdot [h - h_{T0} + T_0 \cdot (s - s_{T0})]$
Mass flow (Mechanical exergy)	$p \neq p_0$ at $T = T_0$	1	$\dot{m} \cdot [h_{T0} - h_0 - T_0 \cdot (s_{T0} - s_0)]$	$\dot{m} \cdot T_0 \cdot (s_{T0} - s_0)$
Substance flow (nonreactive exergy)	$x_j \neq x_{j,0}$ at $T = T_0$ and $p = p_0$	1	$\dot{m}_j \cdot [h_j - h_{j,0} - T_0 \cdot (s_j - s_{j,0})]$ or $\dot{n}_j \cdot [\bar{h}_j - \bar{h}_{j,0} - T_0 \cdot (\bar{s}_j - \bar{s}_{j,0})]$	$\dot{m}_j \cdot T_0 \cdot (s_j - s_{j,0})$ or $\dot{n}_j \cdot T_0 \cdot (\bar{s}_j - \bar{s}_{j,0})$

Table A.3: Overview on transformation energy and transformability associated with various types of mass and energy flows - part 2

Type of flow	Property of mass flow which is different from the environment	Transformability τ_i	Transformation energy flow E_{n_τ}	compensation heat flow Q_0^*
Pure substance flow (Reactive exergy)	substance is not present in the reference environment	1	$\dot{n}_F \cdot (\bar{h}_{f_F} - T_0 \bar{s}^o{}_F)$ $+ \dot{n}_F \cdot \sum_R \frac{\nu_R}{\nu_F} \cdot (\bar{h}_{f_R} - T_0 \bar{s}^o{}_R)$ $- \dot{n}_F \cdot \sum_P \frac{\nu_P}{\nu_F} \cdot (\bar{h}_{f_P} - T_0 \bar{s}^o{}_P)$	$\dot{n}_F \cdot T_0 \bar{s}^o{}_F$ $+ \dot{n}_F \cdot \sum_R \frac{\nu_R}{\nu_F} \cdot T_0 \bar{s}^o{}_R$ $- \dot{n}_F \cdot \sum_P \frac{\nu_P}{\nu_F} \cdot T_0 \bar{s}^o{}_P$
Mass flow (Physical exergy)	only valid if: $T \geq T_0$ and $p > p_0$	$\dfrac{h - h_0 - T_0 \cdot (s(\mathbf{T},p) - s_0)}{h - h_0 - T_0 \cdot (s(\mathbf{T}_0,p) - s_0)}$	$\dot{m} \cdot [h - h_0 - T_0 \cdot (s(\mathbf{T}_0,p) - s_0)]$	$\dot{m} \cdot T_0 \cdot [s(\mathbf{T}_0,p) - s_0]$
Mass flow (Physical exergy)	only valid if: $T < T_0$ and $p > p_0$	$\dfrac{h(\mathbf{T},p) - h_0 - T_0 \cdot (s - s_0)}{h(\mathbf{T}_0,p) - h_0 - T_0 \cdot (s - s_0)}$	$\dot{m} \cdot [h(\mathbf{T}_0,p) - h_0 - T_0 \cdot (s - s_0)]$	$\dot{m} \cdot [h - h_{T0} + T_0 \cdot (s - s_0)]$

A5 Summary of expressions for effective thermal transformabilities

In Table A.4 the effective thermal transformabilities and the matching effective thermal transformation energies are summarized from the derivations in section 2.5 for quick reference.

Table A.4: General expressions for effective thermal transformabilities

Temperature range	$\dot{E}n_{\tau}^{H}$	τ^{H}
$T_i > T_0$ and $T_e > T_0$	$\dot{m} \cdot [(h_i - h_{T0,i}) - (h_e - h_{T0,e})]$	$1 - T_0 \cdot \dfrac{(s_i - s_{T0,i}) - (s_e - s_{T0,e})}{(h_i - h_{T0,i}) - (h_e - h_{T0,e})}$
$T_i < T_0 < T_e$	$\dot{m} \cdot [-T_0 \cdot (s_i - s_{T0,i}) - (h_e - h_{T0,e})]$	$1 - \dfrac{T_0 \cdot (s_e - s_{T0,e}) + (h_i - h_{T0,i})}{T_0 \cdot (s_i - s_{T0,i}) + (h_e - h_{T0,e})}$
$T_e < T_0 < T_i$	$\dot{m} \cdot [-T_0 \cdot (s_e - s_{T0,e}) - (h_i - h_{T0,i})]$	$1 - \dfrac{T_0 \cdot (s_i - s_{T0,i}) + (h_e - h_{T0,e})}{T_0 \cdot (s_e - s_{T0,e}) + (h_i - h_{T0,i})}$
$T_i < T_0$ and $T_e < T_0$	$-\dot{m} \cdot [T_0 \cdot (s_i - s_{T0,i}) - T_0 \cdot (s_e - s_{T0,e})]$	$1 - \dfrac{(h_i - h_{T0,i}) - (h_e - h_{T0,e})}{T_0 \cdot [(s_i - s_{T0,i}) - (s_e - s_{T0,e})]}$

Appendices

A6 Calculation of ideally required heat transfers

Many processes can be assessed directly based on the definitions and descriptions presented in the main body of this dissertation. However, to allow a better understanding, it appears sensible to briefly discuss the definition of ideally required heat flows for some common processes.

A6.1 Evaluation of refrigeration machines

Refrigeration machines that operate with flows above and below reference temperature are very common processes. Therefore the calculation of the ideally required heat flows should be discussed. The problematic aspect for the evaluation of such processes is the fact that transformation energy and transformability are defined differently for flows above and below reference temperature (see Table A.2 on page 130). Thus, while for refrigeration machines operating completely below reference temperature the heat discharge is associated with an exergy and consequently a transformation energy influx, heat discharged above reference temperature is considered a loss to the process. Following the discussion in subsection 3.2.3 on page 54 ff., it is assumed that the heat discharged from a refrigeration machine operating above and below reference temperature is discharged at reference temperature. The ideally required heat flow $\dot{Q}_{0,e}^{id}$ that such a refrigeration machine must discharge in order to operate can be calculated from the energy balance of the reversible refrigeration machine operating between the considered temperatures as a function of the refrigeration capacity $\dot{Q}_l^H$ and the ideal coefficient of performance COP^{id}:

$$\begin{aligned}\dot{Q}_{0,e}^{id} &= -\dot{Q}_l^H \cdot \left(1 + \frac{1}{COP^{id}}\right) \\ &= -\dot{Q}_l^H \cdot \left(1 + \frac{T_{a,h} - T_{a,l}}{T_{a,l}}\right)\end{aligned}$$

The ideally required heat discharge of such a process is consequently the sum of the heat extracted at the low temperature $\dot{Q}_l^H$ and the minimally required power input, which is a a function of the ideal coefficient of performance. The negative sign is required to indicate that the heat discharge has a different direction than the heat extraction flow. Examples of the application of this approach can be found in Table 4.2 on page 71 and in section 4.4 on page 78, which are based on calculations presented in appendices A11 on page 152 ff. and A13 on page 164 ff..

A6.2 Evaluation of heat exchangers

For all applications, where a mass flow enters at a temperature above reference temperature and exits at a temperature below reference temperature or the other way round an effective transformability of the transformation energy difference can be calculated as explained in section 2.5 on page 43 ff. An

Appendices

example of the application of this approach are the heat exchangers of a vapor-compression cascade refrigeration machine that are discussed in section 4.5 on page 84 based on calculations presented in appendix A14 on page 165.

A6.3 Evaluation of heat engines

Heat engines always operate between at least two thermal reservoirs or heat exchangers that can be considered to be quasi-reservoirs at the appropriate average temperature. A heat engine must thus discharge at least as much heat as a reversible heat engine would have to discharge if provided with a given high temperature heat flow by discharging heat at the low temperature of the process[5]. Thus, using the ideal efficiency of a heat engine $\eta^{EL,id}$ the ideally required heat discharge $\dot{Q}_{0,e}^{id}$ can be calculated as a function of the effective heat input into the heat engine $\dot{Q}_h^H$:

$$\begin{aligned} \dot{Q}_{0,e}^{id} &= \left(1 - \eta^{EL,id}\right) \cdot \dot{Q}_h^H \\ &= \left[1 - \left(1 - \frac{T_{a,l}}{T_{a,h}}\right)\right] \cdot \dot{Q}_h^H \\ &= \left(\frac{T_{a,l}}{T_{a,h}}\right) \cdot \dot{Q}_h^H \end{aligned}$$

Since this flow, although generated at $T_l \geq T_0$, is discharged to the environment, the heat flow quickly assumes reference temperature. Therefore, it is considered a heat discharge at reference temperature. The transformability destruction associated with the temperature decrease of the discharged heat flow is consequently influencing the transformability ratio of the process, indicating improvement potential in process parameters or design. Using the ideally required heat flow in the effective compensation heat flow, the heat engine can be evaluated on a scale of 0 to 100 % for transformation energy efficiency and transformability ratio.

A6.4 Evaluation of compressors and expanders

Like heat engines nonadiabatic expanders and compressors interact with their surroundings by exchange of heat flows, even if reversible processes are considered. The comparison of the output temperature of the real process with the output temperature of a reversible adiabatic process [6], which compresses or expands the considered input flow at input temperature T_i and input pressure p_i to exit pressure p_e at constant specific entropy s_i, can provide insight regarding to necessary thermal interaction of the process with the environment.

The evaluation of compressors with a temperature of the exit flow above the exit temperature of a reversible adiabatic process implies that all thermal energy can potentially exit the process with the product flow so that no additionally heat needs to be discharged from the reversible compressor

---

[5] Were the temperature of the environment considered instead, a negative influence of the parameter choice on transformation energy efficiency would result, which is not desirable.

[6] The exit temperature of a reversible adiabatic process is a function of the pressure of the exiting flow and the specific entropy of the input flow and can be obtained by using an appropriate equation of state in order to obtain $T(p_e, s_i)$.

Appendices

operating between the two pressures considered. All heat losses from such a process are optional and need not to be considered as a summand in effective compensation heat flow. The evaluation of such compressors can thus be performed without having to consider an ideally required heat flow.

In contrast to that, compressors with exit temperatures below the exit temperature of a reversible adiabatic compressor operating between the two considered pressures have to discharge a heat flow even in case of reversible operation. Therefore, it is necessary to calculate an ideally required heat flow.

Ideally required heat flows are defined as flows at reference temperature. For the examples considered in this work the process temperatures fulfill $T_i \leq T_0 \leq T_e$ or $T_e \leq T_0 \leq T_i$. Since with these parameters a heat exchange of process and environment without transformability destruction is theoretically possible, matching reversible processes can be evaluated that compress the input flow to the parameters of the exit flow using a minimum amount of work by discharging heat only at reference temperature. Because these processes have no loss flows apart from the ideally required heat flow $\dot{Q}^{id}_{0,e}$, an expression for this heat flow can be derived based on a combination of the energy and the exergy balance of a reversible compressor operating between the two considered pressures and temperatures. The energy balance can be expressed as a function of the ideally required work flow $\dot{W}^{id}$ and the input (subscript i) and exit (subscript e) enthalpy flows $\dot{H}$:

$$0 = \dot{W}^{id}_i + \dot{H}_i(T_i, p_i) - \dot{H}_e(p_e, T_e) - \dot{Q}^{id}_{0,e} \tag{a.4}$$

The exergy balance can be expressed using physical exergy flows $\dot{E}^{PH}$ and exergy flows associated with work $\dot{E}^W$ as[7]:

$$0 = \dot{E}^W + \dot{E}^{PH}_i - \dot{E}^{PH}_e \tag{a.5}$$

Since $\dot{W}^{id} = \dot{E}^W$ and the exergy flow associated with the ideally required heat flow is zero, the exergy balance can be expressed as:

$$\dot{W}^{id} = -\dot{E}^{PH}_i + \dot{E}^{PH}_e$$

Using this expression together with Equation a.4 the following equation for the ideally required heat flow is obtained:

$$\dot{Q}^{id}_{0,e} = -\dot{E}^{PH}_i + \dot{E}^{PH}_e + \dot{H}_i(T_i, p_i) - \dot{H}_e(p_e, T_e)$$

With the definition of physical exergy from Equation 1.3 the definition of the ideally required heat

---

[7] A separate evaluation of mechanical and thermal exergy is not necessary in order to draw the exergy balance as the different types of exergy are added up in the balance together with their algebraic sign. The use of physical exergy simplifies the derivation in this case.

flow can be expressed as follows:

$$\begin{aligned}\dot{Q}^{id}_{0,e} &= -\left\{\dot{H}\left(T_i,p_i\right) - \dot{H}\left(T_0,p_0\right) - T_0 \cdot \left[\dot{S}\left(T_i,p_i\right) - \dot{S}\cdot\left(T_0,p_0\right)\right]\right\} \\ &+ \dot{H}\left(T_e,p_e\right) - \dot{H}\left(T_0,p_0\right) - T_0 \cdot \left[\dot{S}\left(T_e,p_e\right) - \dot{S}\cdot\left(T_0,p_0\right)\right] \\ &+ \dot{H}_i\left(T_i,p_i\right) - \dot{H}_e\left(p_e,T_e\right) \\ &= T_0 \cdot \left[\dot{S}\left(T_i,p_i\right) - \dot{S}\cdot\left(T_e,p_e\right)\right] \\ &= \dot{m}\cdot T_0 \cdot \left[s\left(T_i,p_i\right) - s\cdot\left(T_e,p_e\right)\right] \end{aligned} \quad (a.6)$$

Thus, using energy and exergy balances the ideally required heat flow from the compressor to the environment can be determined as a function of the specific entropies of input and exit.

Thermodynamically, expanders can be considered reversed compression machines. If the exiting mass flow from an expander has a temperature below the reversible adiabatic exit temperature, the consideration of an ideally required heat flow is not necessary as this temperature could have been attained by adiabatic expansion and low temperature heat extraction. If the exit temperature of the mass flow is higher than the exit temperature of an adiabatic reversible expander operating between the same two pressures, the ideally required heat flow $\dot{Q}^{id}_{0,i}$ into an expander has to be accounted for in the effective compensation heat flow. Although this is commonly the case, as friction and irreversibilities increase the temperature of the exit flow, while at the same time reducing the generated work flow, it is also possible to model a reversible process that uses a heat flow from the environment to attain the output temperature T_e. The input and exit temperatures are again assumed to fulfil $T_i \leq T_0 \leq T_e$ or $T_e \leq T_0 \leq T_i$, to allow the inflow of heat at reference temperature without transformability destruction. The ideally required heat input from the environment can be calculated in analogy to Equation a.9 based on the energy and exergy balance of a reversible expander. The energy balance of the reversible expander can be drawn as following:

$$0 = -\dot{W}^{id}_e + \dot{Q}^{id}_{0,i} + \dot{H}_i\left(T_i,p_i\right) - \dot{H}_e\left(T_e,p_e\right) \quad (a.7)$$

The exergy balance of a reversible expander can be expressed as:

$$0 = -\dot{E}^W + \dot{E}^{PH}_i - \dot{E}^{PH}_e \quad (a.8)$$

Since $\dot{W}^{id} = \dot{E}^W$ and the exergy flow associated with, the ideally required heat flow is zero the exergy balance can be expressed as:

$$\dot{W}^{id} = \dot{E}^{PH}_i - \dot{E}^{PH}_e$$

Using this expression together with equation a.7, the following equation for the ideally required heat flow is obtained:

$$\dot{Q}^{id}_{0,i} = \dot{E}^{PH}_i - \dot{E}^{PH}_e - \dot{H}_i\left(T_i,p_i\right) + \dot{H}_e\left(p_e,T_e\right)$$

Using the definition of physical exergy from Equation 1.3, the definition of the ideally required heat

Appendices

flow can be expressed as follows:

$$\begin{aligned}
\dot{Q}_{0,i}^{id} &= \left\{ \dot{H}(T_i, p_i) - \dot{H}(T_0, p_0) - T_0 \cdot \left[\dot{S}(T_i, p_i) - \dot{S} \cdot (T_0, p_0) \right] \right\} \\
&\quad - \dot{H}(T_e, p_e) - \dot{H}(T_0, p_0) - T_0 \cdot \left[\dot{S}(T_e, p_e) - \dot{S} \cdot (T_0, p_0) \right] \\
&\quad - \dot{H}_i(T_i, p_i) + \dot{H}_e(p_e, T_e) \\
&= -T_0 \cdot \left[\dot{S}(T_i, p_i) - \dot{S} \cdot (T_e, p_e) \right] \\
&= -\dot{m} \cdot T_0 \cdot [s(T_i, p_i) - s \cdot (T_e, p_e)]
\end{aligned} \quad (a.9)$$

Thus, the definition of the reversible heat inflow into an expander differs from the definition of the ideal heat exit flow of a reversible compressor in Equation a.6 only by the negative sign.

If input and exit temperatures of the compressor or the expander do not fulfill $T_i \leq T_0 \leq T_e$ or $T_e \leq T_0 \leq T_i$, the reversible compression and expansion processes cannot be described by the exergy balances in Equations a.5 and a.8, since heat interactions with the surroundings at reference temperature in such processes unavoidably result in exergy destruction. Only heat flows at temperatures between input and exit temperature can exit or enter the compression or expansion process. If such heat flows were evaluated at reference temperature, the exergy associated with each of the conductive heat flows would have been destroyed. As a consequence, the validity of Equations a.6 and a.9 has to be assessed individually for cases in which a compressor or expander operates strictly above or stricly below reference temperature.

A7 Calculation basics for the evaluation of heat production from combined heat and power processes

The exergy-based fuel attribution to heat from combined heat and power (CHP) process provides a basis for the transformability assessment for heat from combined heat and power. With this attribution method a CHP process can be divided for the evaluation into a subsystem generating heat and a subsystem generating electricity. Figure A.2 shows the resulting CHP process flow chart.

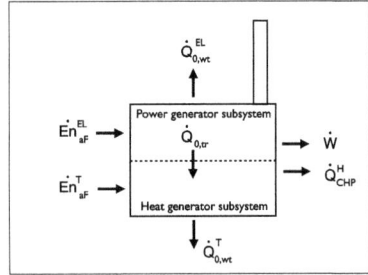

Figure A.2: Flow chart of the separation of a CHP process into subsystems

To fulfill the energy balance the fuel attributed to electricity is the difference between total fuel input

Appendices

and fuel attributed to heat, thus increasing the effective electrical efficiency of CHP plant. Commonly, the fuel attributed to heat from a CHP process is smaller or equal than the heat generated. To fulfill the energy balance of the subsystem generating heat, waste heat from the power process that makes up for the difference between attributed fuel and heat output has to be considered an input into the heat generator subsystem. The waste heat is considered to be heat at reference temperature since it is discharged into the environment if not used by the heat generation subsystem.

An average transformability associated with the useful heat flow from CHP $\tau_{a,CHP}^{H}$ can be calculated as a function of the effective thermal exergy flow $\dot{E}_{aF}^{H}$ associated with the exiting heat flow $\dot{Q}_{CHP}^{H}$. The calculation can be based on Equation 3.7 if the necessary simplifying assumptions are made[8] and can be expressed as a function of the effective thermal transformability of the heat flow τ^{H} and thermal (superscript T) and electrical (superscript EL) efficiency:

$$\begin{aligned} \tau_{a,CHP}^{H} &= \frac{\dot{E}_{aF}^{H}}{\dot{Q}_{CHP}^{H}} \\ &= \frac{f_{aF}^{H} \cdot \dot{E}n_{\tau,CHP,F}}{\dot{Q}_{CHP}^{H}} \\ &= \frac{f_{aF}^{H}}{\eta_{CHP}^{T}} \\ &= \frac{\tau^{H}}{(\eta^{T} \cdot \tau^{H} + \eta^{EL})} \end{aligned} \qquad (a.10)$$

It is assumed that only the net waste heat input into the heat generator subsystem $\Delta \dot{Q}_{0,i}$ is considered in the calculation of the evaluation ratios. It is obtained from the difference of the waste heat flow transferred from the power generator subsystem to the heat generator subsystem $\dot{Q}_{0,tr}$ and the waste heat lost from the heat generator subsystem $\dot{Q}_{0,wt}^{T}$. This assumption is following the rules laid out in subsection 3.2.3 on page 54 ff. for dealing with heat at reference temperature, which states that only net heat flows at a given reference temperature should be considered. Thus, the thermal transformation energy efficiency that allows the separate consideration of the heat generator subsystem can be obtained based on the flowchart in Figure 3.7 from the following equation:

$$\eta_{\tau}^{T} = \frac{\dot{Q}_{CHP}^{H}}{\dot{E}n_{aF}^{H} + \Delta \dot{Q}_{i}^{*}} = 100\,\% \qquad (a.11)$$

The heat generator is thus considered to operate energetically ideal, while its transformability ratio characterizes the subsystem process. Complementing thermal transformation energy efficiency of the heat generating subsystem, the electrical transformation energy efficiency η_{τ}^{EL} of the power generating subsystem is calculated as a function function of the work flow $\dot{W}$, the combustible fuel flow attributed to the heat flow $\dot{E}n_{aF}$ and the total energy flow of the combustible fuel to the combined process

---

[8]To allow the use of this equation, it is assumed that chemical exergy equals the higher heating value of a flow of combustible fuel and that energy efficiencies are given in relation the higher heating value of the fuel.

$\dot{E}n_{cF}$ as:

$$\begin{aligned}\eta_\tau^{EL} &= \frac{\dot{W}}{\dot{E}n_{aF}^{EL}} \\ &= \frac{\dot{W}}{\dot{E}n_{cF} - \dot{E}n_{aF}} \\ &= \frac{\frac{\dot{W}}{\dot{E}n_{cF}}}{\frac{\dot{E}n_{cF}}{\dot{E}n_{cF}} - \frac{\dot{E}n_{aF}}{\dot{E}n_{cF}}} \\ &= \frac{\eta_{CHP}^{EL}}{1 - f_{aF}^{H}} \end{aligned} \qquad (a.12)$$

The average transformabilities associated with fuel and products of the power generation subsystem are equal. Consequently, the transformability ratio of this subsystem always equals 100 %, thus leaving the electrical transformation energy efficiency as the central measure of characterization for the power generating subsystem. It is noteworthy that the electrical transformation energy efficiency of the power generating subsystem is larger than the electrical energy efficiency of the combined heat and power process, since a share of the transformability destruction and transformation energy loss is attributed to the produced heat.

This approach can be extended to evaluate the output of a district heating network in terms of transformation energy and average transformability. See appendix A7.1 ff. for an extensive discussion.

A7.1 Calculation of the average transformability associated with heat from CHP delivered by district heating

For the operation of a district heating (DH) network a certain amount of energy, usually electrical power, is needed to compensate for pressure losses in the pipes. This additional power input has also to be considered when evaluating a district heating system. A straightforward approach is to consider the combustible fuel that is required for the electricity production as another input into the supply system and considering the average transformability of the heat separately from the transformability associated with the combustible fuel input for auxiliary power. However, it is also possible to calculate a total average transformability for the whole transformation energy (CHP heat and fuel for pump electricity). This can be useful if transformation energy of CHP is to be compared with other average transformation energies as a whole, such as has been applied for the basic calculations for the average transformability of district heat in section 4.6 on page 88 ff.

The transformability τ of the auxiliary energy (subscript ax) is 100 % while the transformation energy flow $\dot{E}n_{\tau,CHP}$ associated with the heat input into the district heating system $\dot{Q}_{CHP}^H$ is a function of the required heat flow $\dot{Q}_{rq}$ and the transport efficiency η_{tr}, which is defined as the ratio of heat output from the transport system to the heat input into the transport system.

$$\dot{E}n_{\tau,CHP}^H = \frac{\dot{Q}_{rq}}{\eta_{tr}} = \dot{Q}_{CHP}^H$$

Appendices

Since for conductive heat flows above reference temperature transformation energy equals energy, the discussion can be simplified.

The auxiliary energy can be expressed using an auxiliary energy factor f_{ax} that relates the auxiliary electrical power for the pumps $\dot{W}_{ax}$ in the DH net to the heat generated by the CHP plant. Since the factor relates the auxiliary power to district heat, but the relevant input into the supply system is the fuel input into the power generator it is important to consider the appropriate electrical efficiency η_{ax}^{EL}, depending on the source of the power generation for auxiliary energy. If the auxiliary power is generated by the CHP plant, the relevant efficiency is given by Equation a.12.

The auxiliary fuel factor is defined as:

$$f_{ax} = \frac{\dot{W}_{ax}}{\dot{Q}_{CHP}^H}$$

Using this factor, the total input of combustible fuel for auxiliary energy $\dot{En}_{cF,ax}^{TO}$ can be calculated as a function of the energy loss in the fuel to power conversion system $\dot{En}_{ax}^L$ and consequently the electrical efficiency η_{ax}^{EL} of this fuel to power conversion system:

$$\begin{aligned}
\dot{En}_{cF,ax}^{TO} &= \dot{W}_{ax} + \dot{En}_{ax}^L \\
&= \frac{\dot{W}_{ax}}{\eta_{ax}^{EL}} \\
&= \frac{f_{ax} \cdot \dot{Q}_{rq}}{\eta_{ax}^{EL} \cdot \eta_{tr}}
\end{aligned}$$

The effective thermal transformation energy factor f^H, which is defined by Equation 2.36 for the calculation of average transformabilities, can be expressed as follows:

$$\begin{aligned}
f^H &= \frac{\dot{Q}_{CHP}^H}{\dot{Q}_{CHP}^H + \dot{En}_{F,ax}^{TO}} \\
&= \frac{\dfrac{\dot{Q}_{rq}}{\eta_{tr}}}{\dfrac{\dot{Q}_{rq}}{\eta_{tr}} + \dfrac{f_{ax} \cdot \dot{Q}_{rq}}{\eta_{ax}^{EL} \cdot \eta_{tr}}} \\
&= \frac{1}{1 + \dfrac{f_{ax}}{\eta_{ax}^{EL}}} \\
&= \frac{\eta_{ax}^{EL}}{f_{ax} + \eta_{ax}^{EL}}
\end{aligned} \qquad (a.13)$$

The auxiliary transformation energy factor $f_{\tau,ax}$ can be calculated analogously as:

$$f_{\tau,ax} = \frac{\dot{En}_{F,ax}^{TO}}{\dot{Q}_{CHP}^{H} + \dot{En}_{F,ax}^{TO}}$$

$$= \frac{\frac{f_{ax} \cdot \dot{Q}_{rq}}{\eta_{ax}^{EL} \cdot \eta_{tr}}}{\frac{\dot{Q}_{rq}}{\eta_{tr}} + \frac{f_{ax} \cdot \dot{Q}_{rq}}{\eta_{ax}^{EL} \cdot \eta_{tr}}}$$

$$= \frac{\frac{f_{ax}}{\eta_{ax}^{EL}}}{1 + \frac{f_{ax}}{\eta_{ax}^{EL}}}$$

$$= \frac{\frac{f_{ax}}{\eta_{ax}^{EL}}}{\frac{\eta_{ax}^{EL} + f_{ax}}{\eta_{ax}^{EL}}}$$

$$= \frac{f_{ax}}{\eta_{ax}^{EL} + f_{ax}} \tag{a.14}$$

As a consequence, the average transformability associated with district heat $\tau_{a,DH}$ can be calculated based on Equations 3.5, a.13 and a.14 as:

$$\tau_{a,DH} = f^{H} \cdot \tau_{CHP}^{H} + f_{\tau,ax} \cdot \tau_{ax}$$
$$= \frac{\eta_{ax}^{EL}}{f_{ax} + \eta_{ax}^{EL}} \cdot \frac{\tau^{H}}{(\eta^{T} \cdot \tau^{H} + \eta^{EL})} + \frac{f_{ax}}{\eta_{ax}^{EL} + f_{ax}}$$

The total transformation energy to which this average transformability is associated can be expressed as a function of the heat required by the supply target as:

$$\dot{En}_{\tau}^{TO} = \frac{\dot{Q}_{rq}}{\eta_{tr}} + \frac{f_{ax} \cdot \dot{Q}_{rq}}{\eta_{ax}^{EL} \cdot \eta_{tr}}$$

The total average transformability for CHP heat is valid under the assumption that all pump power is simply lost and not recovered in form of useful heat in the district heating water. Further assumptions include a consideration of thermal and chemical transformation energy only and the assumption of the equivalence of the higher heating value of a fuel and the chemical exergy assoiciated with it.

A8 Calculating evaluation ratios for the evaluation of heat and cold supply systems

The rules laid down for the definition of a suitable evaluation boundary in section 3.3 on page 59 ff. require the definition of a required demand, which is equal for all systems to be compared. For a basic transformability evaluation of heat and cold supply systems, it is sufficient to model the required

Appendices

thermal demand with conductive heat flows at a given temperature. Thus, a general transformation energy efficiency η_τ of a heating system can be defined as a function of the transformation energy associated with the required heat flow $\dot{E}n_{\tau,rq}^Q$ and the transformation energy flow associated with the effective thermal energy input flows $\dot{E}n_{\tau,i}^H$ and the chemical transformation energy input flows $\dot{E}n_{\tau,i}^{CH}$:

$$\eta_\tau = \frac{\dot{E}n_{\tau,P}}{\dot{E}n_{\tau,F}} = \frac{\dot{E}n_{\tau,rq}^Q}{\sum_i \dot{E}n_{\tau,i}^H + \sum_i \dot{E}n_{\tau,i}^{CH}}$$

Assuming that the higher heating value HHV is equal to the chemical transformation energy and using heat flows $\dot{Q}$ instead of transformation energy flows based on Equations 2.4 and 2.33 this ratio can be expressed as:

$$\eta_\tau = \eta = \frac{\dot{Q}_{rq}}{\sum \dot{Q}^H + \sum H\dot{H}V_{cF,i}}$$

The transformability ratio ξ can most easily be calculated as the ratio of exergetic efficiency ε to transformation energy efficiency. If the exergetic efficiency is defined as a ratio of exergy flows $\dot{E}$:

$$\varepsilon = \frac{\dot{E}_{rq}^Q}{\sum \dot{E}_i^H + \sum \dot{E}_i^{CH}}$$

the transformability ratio can be defined as a function of average transformabilities τ_a of product (subscript P) and fuel (subscript F) as defined in Equation 2.35:

$$\xi = \frac{\varepsilon}{\eta_\tau} = \frac{\dfrac{\dot{E}_{rq}^Q}{\dot{E}n_{\tau,rq}^Q}}{\dfrac{\sum \dot{E}_i^H + \sum \dot{E}_i^{CH}}{\sum \dot{E}n_{\tau,i}^H + \sum \dot{E}n_{\tau,i}^{CH}}} = \frac{\tau_{a,P}}{\tau_{a,F}}$$

The definition of the transformation energy efficiency, the exergetic efficiency and the transformability ratio for cooling supply systems are in principle equal to the definitions used for heat supply systems. The only significant difference is the necessity to consider the effective compensation heat flow in the transformation energy efficiency and as a consequence also in the transformability ratio. The effective compensation heat flow can be calculated according to Equation 2.1 and has to be considered either in the denominator of the transformation energy efficiency if greater zero or in its numerator if lesser zero.

A9 Summary of the transformability assessment method for energy supply technologies

1. Define the target supply parameters (like temperature and required transformation energy) that should be kept constant, compensating for unwanted effects in the supply target (such as heat loss).

Appendices

2. Define or calculate the transformability and transformation energy demand.
3. Define the cross-comparable system boundaries for each supply technology, so that:
   - the required transformation energy crosses the system boundary entering the supply target
   - the primary energy converter is within the boundary if the input energy flow is a nonwaste storable energy form (combustible fuels)
   - the connection to the primary energy converter is within the boundary if the input energy flow is a nonstorable energy flow (sunlight, wind)
     - Afterwards, technology-specific subsystems are defined which transform nonstorable energy (such as wind or solar radiation into a storable energy form (potential energy or heat))
   - the attributed fuel and the waste heat flow enter the system boundary if the energy is supplied by a cogeneration system
   - waste heat flows from industrial processes enter the system
     - Waste heat flows are flows that are normally discharged to the environment. If these flows are used without having an influence on the original process, they are considered at the output of the waste-heat generator. If the use of waste heat significantly affects the primary output, the process has to be considered a cogeneration process and evaluated accordingly.
4. Calculate transformation energies, compensation heat flows and ideal heat flows as well as exergies for of all relevant flows using Tables A.18 on page 168 and A.19 on page 169
5. Calculate transformation energy efficiency and transformability ratio (using the exergetic efficiency).

A10 Calculations for comparative evaluation of supply systems for domestic heating

In this section the calculations which lay the basis for the results presented in Table 4.1 on page 68 are presented[9].

The common supply target of the compared heating systems is a single family house with a transmissive heat loss of 12 kW to the environment through the walls and windows. All heat losses from the target volume (the house) are considered to be transmissive. The balance boundaries are set in such a way that 12 kW conductive heat at room temperature have to be supplied by the heating system to maintain room temperature. The reference temperature is set to 1,85 °C (275 K). Calculations are based on conductive heat flows and fuel mass flows. The difference between the higher heating value

---

[9]All calculations have been performed using nonrounded values. However, to be able to display intermediate results in this section these results have been rounded to two decimal places. Using these rounded values of the intermediate results as a basis for calculating the final results, might provide slightly different final results.

Appendices

of a fuel and chemical exergy transformation energy is neglected. Pressure losses and consequently mechanical exergy and transformation energy in all components of the energy supply system are neglected, since pressure losses are usually low compared to thermal transformation energy losses and transformability destruction. Most of the basic data is assumed according to common engineering experience instead of having been researched for specific real examples, since the purpose of the assessment is to demonstrate the viability of the comparative transformability assessment method and not to perform a comparative assessment of real processes.

In order to minimize the steps of calculation the transformability ratio ξ is calculated as the ratio of exergetic efficiency and transformation energy efficiency instead of calculating ξ directly from average in- and output transformabilities.

Table A.5 shows the general assumptions underlying the evaluation of the four example processes. Based on these assumptions the four processes are evaluated.

Flow charts of the considered supply systems are provided for every system. Numbered flows indicate flows that do not have to be explicitly calculated to allow an assessment, while flows labeled according to the general nomenclature have to be calculated to allow evaluation.

Table A.5: General assumptions for the comparative evaluation of heat supply systems

Data	Symbol	Value	Units	Comment
Required heat input	$\dot{Q}_r$	12,00	kW	All compared heating systems supply this amount of heat at room temperature to the supply target.
Room temperature	T_r	295,00	K	
Room pressure	p_r	101,32	kPa	
Reference temperature	T_0	275,00	K	
Reference pressure	p_0	101,32	kPa	
Average efficiency of power generation in Germany 2005	η_a^{EL}	41 %		(Machat and Werner, 2007)

The fuel-related coefficients of performance $COP_{a,cF}$ are calculated for all considered technologies as functions of the transformation energy flow associated with combustible fuels $\dot{En}_{\tau,cF}$ as:

$$COP_{a,cF} = \frac{\dot{Q}_r}{\dot{En}_{\tau,cF}}$$

A10.1 Boiler

Table A.6 summarizes the specific assumptions that are required additionally to the general assumptions in Table A.5 to evaluate a heat supply system based on a natural gas condensing boiler.

Table A.6: Specific assumptions for the evaluation of a heat supply system based on a condensing boiler

Data	Symbol	Value	Comment
Average annual efficiency of the condensing boiler	η^T	$95,00\,\%$	Ratio of heat output to the higher heating value of the fuel
Share of auxiliary work flow for boiler operation in relation to generated heat	f_{ax}	$0,50\,\%$	Required to operate pumps and auxiliary equipment

Figure A.3 shows the flow charts of the considered heat supply system. Flow 1 indicates the losses in the power generator that generates the auxiliary power, which is symbolized by flow 2. Flow 3 is the heat loss from the supply target to the environment.

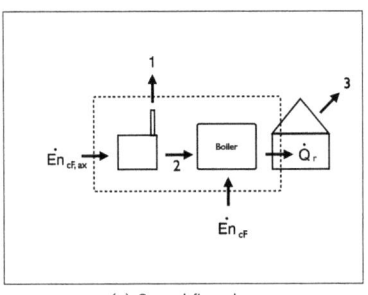
(a) General flow chart

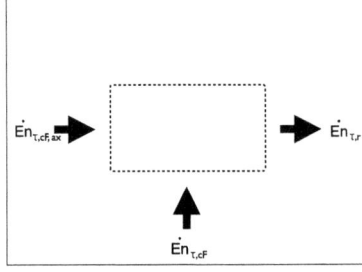
(b) Transformation energy flow chart

Figure A.3: Flow charts of a heat supply system based on a gas condensing boiler

Since chemical exergy and chemical transformation energy are assumed to be equal to the higher heating value the transformation energy of the fuel can be calculated as:

$$\dot{En}_{T,cF} = \frac{\dot{Q}_r}{\eta^T} = 12,63\,kW$$

Similarly, the fuel required to generate the auxiliary power can be calculated as:

$$\dot{En}_{T,cF,ax} = \frac{\dot{Q}_r \cdot f_{ax}}{\eta_a^{EL}} = 0,15\,kW$$

Thus, the transformation energy efficiency of the heat supply system is given as:

$$\eta_T = \frac{\dot{En}_{T,r}}{\dot{En}_{T,cF} + \dot{En}_{T,cF,ax}} = 93,86\,\%$$

Appendices

The exergetic efficiency is given by:

$$\varepsilon = \frac{\dot{E}_r}{\dot{E}_{cF} + \dot{E}_{cF,ax}} = 6,44\,\%$$

The results of the heat supply system evaluation, are summarized in Table 4.1.

A10.2 Heat from a geothermal heat source

The second considered supply system is based on heat obtained from a geothermal heat source and transported to the supply target by a district heating network.

Table A.7: Specific assumptions for the evaluation of a heat supply system based on geothermal district heating

Data	Symbol	Value	Units	Comment
Average efficiency of the transport of hot water from geothermal source to the supply target	η_{tr}	$90,00\,\%$		Ratio of heat output to supply target to output from geothermal source
Share of auxiliary power to pump up the water from the aquifer, and to distribute the hot water to the households	f_{ax}	$3\,\%$		Required to operate pumps and auxiliary equipment In relation to the required heat
Average temperature of heat transfer from geothermal source	T_a	$337,93$	K	The heat transfer from ground occurs on a length of pipe over which heat is transferred at temperatures from 50 °C to 80 °C . The temperature is the average logarithmic mean temperature.

Figure A.4 shows the flow charts for the geothermal heat supply system. Flow 1 designates the heat loss from the power plant that provides the auxiliary energy. Flow 2 represents the auxiliary power provided, while flow 3 is the heat loss of the building that has to be compensated for.

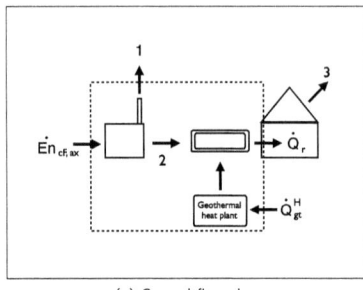
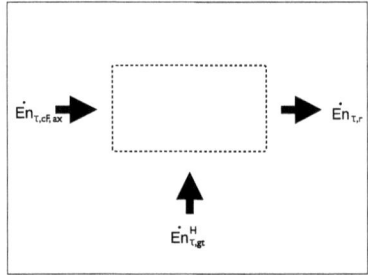

(a) General flow chart   (b) Transformation energy flow chart

Figure A.4: Flow charts of a heat supply system based on a geothermal source

Since chemical exergy and chemical transformation energy are assumed to be equal to the higher heating value, the transformation energy of the fuel required to provide the auxiliary power can be calculated as :

$$\dot{En}_{T,cF,ax} = \frac{f_{ax} \cdot \dot{Q}_r}{\eta_a^{EL}} = 0,98\,kW$$

The heat to be extracted from the geothermal source $\dot{Q}_{gt}$ to provide heat to the supply target equals:

$$\dot{Q}_{gt}^H = \frac{\dot{Q}_r}{\eta_{tr}} = 13,33\,kW$$

Thus, the transformation energy efficiency of the heat supply system is given as:

$$\eta_T = \frac{\dot{En}_{T,r}}{\dot{En}_{T,gt}^H + \dot{En}_{T,cF,ax}} = 83,86\,\%$$

The exergetic efficiency is given by:

$$\varepsilon = \frac{\dot{E}_r}{\dot{E}_{gt}^H + \dot{E}_{cF,ax}} = 23,52\,\%$$

ξ^{mx} is calculated using the equations from this subsection but with the following assumptions: $\eta^{EL,mx} = 100\,\%$, $\eta_{tr}^{mx} = 100\,\%$ and $f_{ax} = 0\,\%$, which results in $\eta_T^{mx} = 100\,\%$ and $\varepsilon^{mx} = \xi^{mx} = 36,41\,\%$.

A10.3 Ground-source heat pump

The ground-source heat pump is the third technology chosen for the exemplary comparison. Table A.8 summarizes the specific assumptions required for an evaluation of the considered supply system.

Appendices

Table A.8: Specific assumptions for the evaluation of a heat supply system based on an electrical ground-source heat pump

Data	Symbol	Value	Units	Comment
Average temperature at the condenser	$T_{a,h}$	310,00	K	Condensation temperature is a function of the used working fluid and its pressure, therefore it is independent of reference temperature.
Average temperature at the evaporator	$T_{a,l}$	283,15	K	10 °C = average temperature of the ground
Ratio of average annual COP to ideal COP	$\dfrac{COP_a}{COP^{id}}$	0,3		$COP^{id} = \dfrac{T_{a,h}}{T_{a,h} - T_{a,l}}$ Moran and Shapiro (2007) This factor was chosen in such a way to approximately result in an annual average COP as given by Klenner (2008).

In Figure A.5 flow charts for the heat supply system are shown. Like in the previous example of the supply system with the geothermal heat supply, flow 1 equals the losses from the condensing power plant that generates flow 2, the power required to operate the compression heat pump. Flow 3 is the heat flow that is lost from the supply target that has to be compensated for by the supply system.

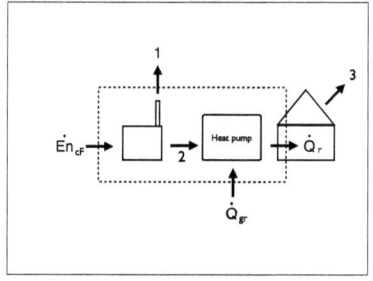
(a) General flow chart

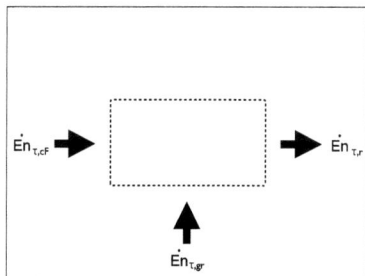
(b) Transformation energy flow chart

Figure A.5: Flow charts of a heat supply system based on a heat pump

The average coefficient of performance can be obtained based on these assumptions as:

$$COP_a = \frac{T_{a,h}}{T_{a,h} - T_{a,l}} \cdot 0,3 = 3,46$$

The annual average COP related to the fuel input is calculated as:

$$COP_{a,cF} = COP_a \cdot \eta_a^{EL} = 1,42$$

The combustible fuel input required to operate the heat pump is therefore:

$$\dot{E}n_{cF} = \dot{E}n_{\tau,cF} = \frac{\dot{Q}_r}{COP_{a,cF}} = 8,45\,kW$$

The heat extracted from the ground $\dot{Q}_{gr}$ is calculated by the energy balance under the assumption of an adiabatic heat pump, which is discharging all heat over the condenser as:

$$\dot{Q}_{gr} = \dot{Q}_r \cdot (1 - \frac{1}{COP_a}) = 8,54\,kW$$

Transformation energy efficiency of the considered supply system can thus be expressed as:

$$\eta_\tau = \frac{\dot{E}n_{\tau,r}}{\dot{E}n_{\tau,cF} + \dot{Q}_{gr}} = 70,65\,\%$$

While exergetic efficiency is calculated as:

$$\varepsilon = \frac{\dot{E}_r}{\dot{E}_{cF} + \dot{E}_{gr}} = 9,36\,\%$$

ξ^{mx} is calculated using the equations from this subsection but with the following assumptions: $\eta^{EL,mx} = 100\,\%$, $COP_a = COP^{id} = 11,55$ which results in $\eta_\tau^{mx} = 100\,\%$ and $\varepsilon^{mx} = \xi^{mx} = 60,05\,\%$. The assessment results are summarized in Table 4.1 on page 68.

A10.4 Block Combined Heat and Power plant

The final exemplary heat supply system is a heat supply system in which the heat originates from a block heat and power plant. See Table A.9.

Appendices

Table A.9: Specific assumptions for the evaluation of a heat supply system based on a block heat and power plant

Data	Symbol	Value	Units	Comment
Average temperature of heat flow from district heating water	$T_{a,DH}$	336,19	K	Logarithmic mean temperature of the forward flow temperature of 85 °C and the return flow temperature of 42 °C
Average annual thermal efficiency of the block CHP plant	η_a^T	49 %		
Average annual electrical efficiency of the block CHP plant	η_a^{EL}	36 %		
Average annual transport efficiency of the district heating network	η_{tr}	85 %		
Share of auxiliary power required for the operation of the district heating network	f_{ax}	2 %		Required to operate pumps and auxiliary equipment. In relation to the heat provided from the CHP plant.

Figure A.6 shows the flow charts for the heat supply system considered. The assessment of heat from combined heat and power is based on the fuel attribution approach discussed in section 3.4 on page 62 ff. Consequently, the total fuel input into the CHP plant (flow 1) is split into a fuel flow attributed to heat $\dot{E}n_{aF}^H$ and one attributed to power. The fuel flow attributed to power is then split into a flow that is used to generate external power (flow 2) and a flow used to generate the auxiliary power to operate the district heating net ($\dot{E}n_{cF,ax}$). Flow 3 symbolizes the losses of the CHP process that are not associated with heat generation or the production of auxiliary power, while flow 4 is the heat loss from the DH network. Flow 5 is the heat loss of the supply target that has to be compensated for.

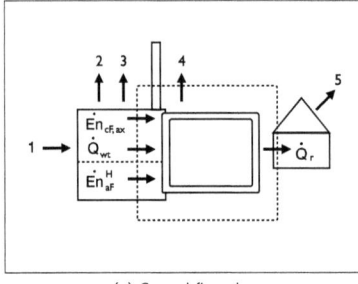 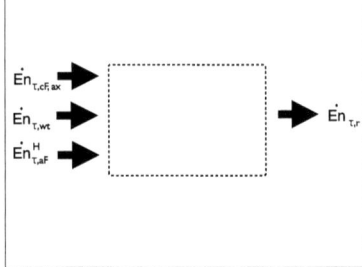

Figure A.6: Flow charts of a heat supply system based on a block heat and power plant

Due to the attribution of a fuel share to the discharged heat, the power generated by CHP has a higher effective transformation energy efficiency than the conventional energy efficiency. The fuel attribution to heat implies that all the exergy of the heat and a share of the losses of the combined process need not to be accounted for in the efficiency of power generation. The electrical part of the CHP process can be considered to be discharging waste heat at reference temperature. This discharged waste heat at reference temperature is symbolized by flow $\dot{Q}_{wt}$, which is directly used as an input into the district heating system. The effective electrical transformation energy efficiency can be calculated using the fuel attribution factor $f_{aF} = 20,10\,\%$ (calculated from Equation 3.5) as:

$$\eta_\tau^{EL} = \frac{\eta_a^{EL}}{1 - f_{aF}^H} = 44,66\,\%$$

Using this efficiency, the fuel input associated with the auxiliary power can be calculated as:

$$\dot{En}_{cF,ax} = \frac{\dot{Q}_r}{\eta_{tr} \cdot \eta_\tau^{EL}} \cdot f_{ax} = 0,63\,kW$$

The fuel attributed to district heating can consequently be expressed as:

$$\dot{En}_{aF}^H = \frac{\dot{Q}_r}{\eta_{tr} \cdot \eta_a^T} \cdot f_{aF} = 5,75\,kW$$

Assuming that $\dot{En}_{F,ax}$ is lost fully with flow 4, the waste heat at reference temperature that the electrical power generation provides freely to the district heating network can be calculated from the energy balance as:

$$\dot{Q}_{wt} = \frac{\dot{Q}_r}{\eta_{tr}} - \dot{En}_{aF}^H = 8,36\,kW$$

In equation a.11 it has been shown that the sum of the effective thermal transformation energy flow associated with the attributed fuel $\dot{En}_{\tau,CHP}^H$ and the net waste heat input $\Delta\dot{Q}_{wt,i}^*$ equals the heat

Appendices

flow from the CHP plant. Thus, the transformation energy efficiency of the considered supply system can be expressed as:

$$\eta_\tau = \frac{\dot{En}_{\tau,r}}{\dot{En}_{aF}^H + \dot{En}_{cF,ax} + \dot{Q}_{wt}} = 81,36\,\%$$

, while exergetic efficiency is calculated as:

$$\varepsilon = \frac{\dot{E}_r}{\dot{E}_{aF}^H + \dot{E}_{cF,ax}} = 12,74\,\%$$

ξ^{mx} is calculated using the equations from this subsection but with the following assumptions: $\eta_{CHP}^{EL,mx} = \eta_{CHP}^{EL,id} = 1 - \frac{336,19\,K}{1350\,K} = 79,29\,\%$, $\eta^{TO} = 100\,\%$ and $\eta_{tr} = 100\,\%$, which results in $\eta_\tau^{mx} = 100\,\%$ and $\varepsilon^{mx} = \xi^{mx} = 30,94\,\%$.

A11 Calculations for comparative evaluation of supply systems for domestic cooling

The following examples for a comparative evaluation of supply systems for domestic cooling are based on the approach to comparative assessment discussed in chapter 3 on page 48 ff.

The common supply target of the compared cooling systems is a single family house, with a transmissive heat influx of 5 kW from the environment through the walls and windows. The balance boundaries are set in such a way that 5 kW conductive heat at room temperature have to be extracted by the cooling system to maintain room temperature. The reference temperature is set to 36,85 °C (310 K). Calculations are based on conductive and effective thermal heat flows and combustible fuel flows. The difference between the higher heating value of a fuel and the associated chemical exergy and transformation energy is neglected. Pressure losses in all components of the energy supply system are neglected as they are usually low compared to thermal transformation energy losses and transformability destruction. Most of the basic data is assumed according to common engineering experience, instead of researched for specific real examples, since the purpose of the assessment is to demonstrate the viability of the comparative transformability assessment and not to perform a comparative assessment of real processes.

Discharge heat flows at temperatures above reference temperature are calculated from the energy balance. All heat discharged through condensers and all heat lost otherwise is considered as being discharged at reference temperature, which implies a total destruction of thermal transformability of heat flows above reference temperature for cooling systems.

All transformabilities, transformation energy flows and compensation heat flows are calculated according to the equations in Tables A.2 on page 130 and on page 131.

In order to minimize the steps of calculation the transformability ratio is calculated as the ratio of exergetic efficiency and transformation energy efficiency instead of calculating ξ directly using average in- and output transformabilities.

Appendices

In addition to transformation energy efficiency, real and maximum transformability ratio and exergetic efficiency the annual average fuel-related coefficient of performance is given, which can be calculated for every supply system from:

$$COP_{a,cF} = \frac{\dot{Q}_r}{\dot{E}n_{\tau,cF}}$$

Table A.10 shows the general assumptions underlying the evaluation of the three example processes. Based on these assumptions the processes are evaluated.

Table A.10: General assumptions for the comparative evaluation of cooling systems

Data	Symbol	Value	Units	Comment
Required heat extraction	$\dot{Q}_r$	5,00	kW	All compared cooling systems extract this heat flow at room temperature from the supply target.
Room temperature	T_r	295,00	K	
Room pressure	p_r	101,32	kPa	
Reference temperature	T_0	310,00	K	
Reference pressure	p_0	101,32	kPa	
Average efficiency of power generation in Germany 2005	η_a^{EL}	41 %		(Machat and Werner, 2007)

Flow charts of the considered supply systems are provided for every supply system. Numbered flows indicate flows that do not have to be explicitly calculated to allow an assessment, while flows labeled according to the general nomenclature have to be calculated to allow evaluation.

A11.1 Compression refrigeration machine

Figures A.7 and A.8 show the flow charts on which the evaluation of the compression refrigeration machine is based, while Table A.11 shows the underlying process-specific assumptions.

Appendices

Table A.11: Specific assumptions for the evaluation of a cooling supply system based on a compression refrigeration machine

Data	Symbol	Value	Units	Comment
Average temperature at the condenser	$T_{a,h}$	325,00	K	Condensation temperature is a function of the used working fluid and its pressure. It is therefore independent of reference temperature.
Average temperature at the evaporator	$T_{a,l}$	282,15	K	9 °C as a mean temperature between the inflow at 6 °C and the exit flow at approximately 12 °C
Ratio of average annual to ideal COP	$\dfrac{COP_a}{COP^{id}}$	0,3		$COP^{id} = \dfrac{T_{a,l}}{T_{a,h} - T_{a,l}}$ Moran and Shapiro (2007)This factor was chosen equal to the factor used for the evaluation of the compression heat pump in section A10 on page 143.

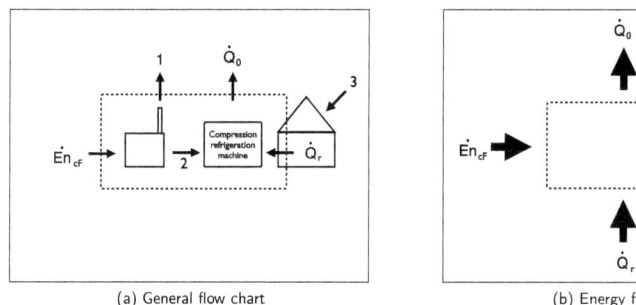

(a) General flow chart   (b) Energy flow chart

Figure A.7: Flow charts of a cold supply system based on a compression refrigeration machine - part 1

In Figure A.7 part (a) flow 1 designates the losses from the electricity generation power plant, flow 2 the transferred electricity and flow 3 the heat influx from the environment into the considered supply target which needs to be compensated for. $\dot{Q}_0$ indicates the heat discharge from the compression refrigeration machine evaluated at reference temperature.

The average coefficient of performance can be obtained based on the assumptions as:

$$COP_a = \frac{T_{a,l}}{T_{a,h} - T_{a,l}} \cdot 0,3 = 1,98$$

The annual average fuel-related COP is calculated as:

$$COP_{a,cF} = COP_a \cdot \eta_a^{EL} = 0,81$$

Under the made assumptions, the combustible fuel input $\dot{E}n_{cF}$ required to operate the refrigeration machine equals the transformation energy flow $\dot{E}n_{\tau,cF}$ it is associated with and can be calculated as a function of the heat extracted from the room $\dot{Q}_r$ as:

$$\dot{E}n_{cF} = \dot{E}n_{\tau,cF} = \frac{\dot{Q}_r}{COP_{cF,a}} = 6,17\,kW$$

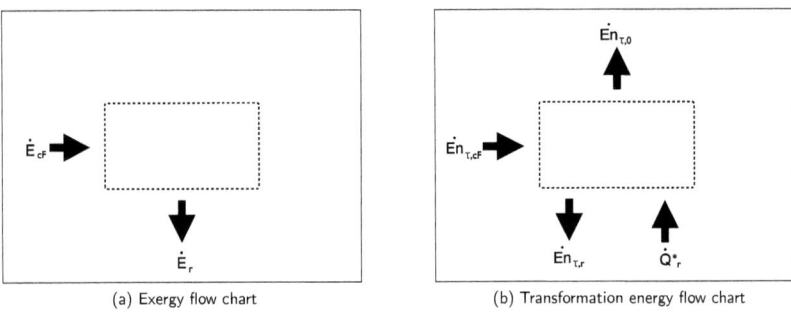

(a) Exergy flow chart (b) Transformation energy flow chart

Figure A.8: Flow charts of a cold supply system based on a compression refrigeration machine - part 2

Although the effective heat output at the condenser is at $T_{a,h} > T_0$ the transformability of the discharged heat is considered to be destroyed. Therefore, the ideally required heat output $\dot{Q}_0^{id}$ from the refrigeration machine is regarded as a heat output at reference temperature T_0 and calculated from the energy balance as:

$$\dot{Q}_0^{id} = \dot{Q}_r + \frac{\dot{Q}_r}{COP^{id}} = 5,76\,kW$$

$\dot{Q}_0^{id}$ is a part of the total heat output of the compression refrigeration machine, which is labeled $\dot{Q}_0$

Appendices

in the flowcharts. Using the compensation heat flow:

$$\dot{Q}_r^* = \left(1 + \frac{T_0}{T_r}\right) \cdot \dot{Q}_r = 10,25\,kW$$

the effective compensation heat flow $\Delta \dot{Q}_i^*$ can be calculated as:

$$\Delta \dot{Q}_i^* = \dot{Q}_r^* - \dot{Q}_0^{id} = 4,49\,kW$$

Since the effective compensation heat flow has a positive sign it has to be considered an input and therefore to be included into the denominator of the transformation energy efficiency η_τ. Transformation energy efficiency of the considered refrigeration machine is a function of the required transformation energy $\dot{E}n_{\tau,r}$ and given by[10]:

$$\eta_\tau = \frac{\left|\dot{E}n_{\tau,r}\right|}{\dot{E}n_{\tau,cF} + \Delta \dot{Q}_i^*} = 49,25\,\%$$

, while exergetic efficiency ε is calculated as a ratio of the relevant exergy flows $\dot{E}$ as:

$$\varepsilon = \frac{\left|\dot{E}_r\right|}{\dot{E}_{cF}} = 4,12\,\%$$

The assessment results are summarized in Table 4.2 on page 71. ξ^{mx} is calculated using the equations from this subsection but with the following assumptions: $COP_a = COP^{id} = 1 - \frac{282,15\,K}{325\,K - 282,15\,K} = 6,58$ and $\eta^{EL} = 100\,\%$, which results in $\eta_\tau^{mx} = 100\,\%$ and $\varepsilon^{mx} = \xi^{mx} = 33,48\,\%$.

A11.2 Direct seawater cooling

The second cold supply system under consideration is a direct seawater cooling system, which is represented by the flow charts in Figures A.9 and A.10.

---

[10] Absolute value bars are used for values with a negative sign, since the negative sign of the transformation energy or exergy associated with the heat flow from the room is only relevant in the respective balances as an indicator of flow direction in relation to the considered mass or energy transfer. In the definition of transformation energy efficiency and exergetic efficiency the flow direction of the transformation energy or exergy has already been considered. Therefore, it is necessary to ensure that all summands in product and fuel have a positive sign.

Appendices

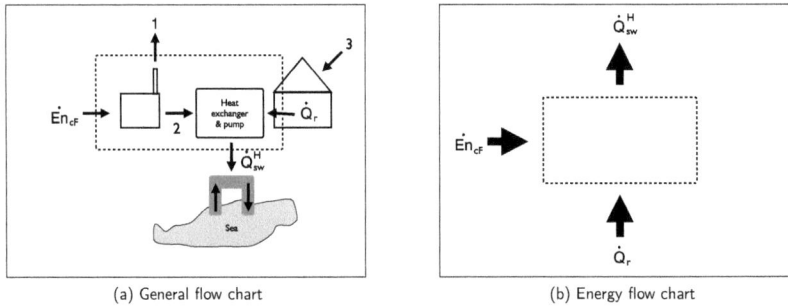

(a) General flow chart (b) Energy flow chart

Figure A.9: Flow charts of a cold supply system based on seawater cooling - part 1

Flow 1 again signifies the energy losses in the power plant, flow 2 stands for the electric power provided to the seawater pump and auxiliary devices, while flow 3 is the heat flow from the environment into the target cooling volume, which has to be compensated for by the cooling system. Specific assumptions for the seawater cooling system are summarized in Table A.12.

Table A.12: Specific assumptions for the evaluation of a seawater cooling system

Data	Symbol	Value	Units	Comment
Average temperature of flow which is provided to cooling system	$T_{a,\,sw}$	282, 15	K	9 °C as a mean temperature between the inflow at 6 °C and the exit flow at approximately 12 °C
Auxiliary power for pump and equipment as a share of the considered cooling capacity	f_{ax}	3 %		

The combustible fuel input into the cooling system equals:

$$\dot{E}n_{cF} = \dot{E}n_{\tau,\,cF} = \frac{\dot{Q}_r}{\eta^{EL}} \cdot f_{ax} = 0,37\,kW$$

Appendices

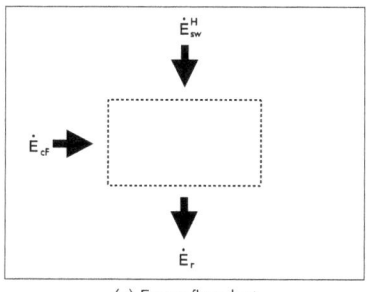
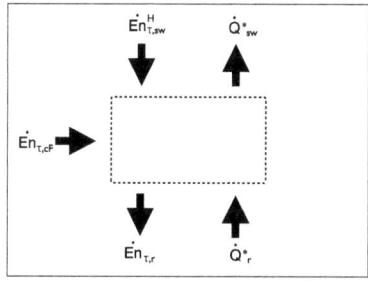

(a) Exergy flow chart   (b) Transformation energy flow chart

Figure A.10: Flow charts of a cold supply system based on direct seawater cooling - part 2

The transformation energy input from the seawater $\dot{En}_{T,sw}$ can be calculated as:

$$\dot{En}^H_{T,sw} = -\dot{Q}_r \cdot \frac{T_0}{T_{a,sw}} = -5,49\,kW$$

The effective compensation heat flow input is given by:

$$\Delta \dot{Q}^*_i = \dot{Q}^*_r - \dot{Q}^*_{sw} = -0,24\,kW$$

Thus, the effective compensation heat flow is an output and has to be considered in the numerator of the transformation energy efficiency.

The transformation energy efficiency for the seawater cooling system can be expressed as:

$$\eta_\tau = \frac{\left|\dot{En}_{T,r}\right| + \left|\Delta \dot{Q}^*_i\right|}{\dot{En}_{T,cF} + \left|\dot{En}^H_{T,sw}\right|} = 93,76\,\%$$

, while exergetic efficiency is calculated as:

$$\varepsilon = \frac{\left|\dot{E}_r\right|}{\dot{E}_{cF} + \left|\dot{E}^H_{sw}\right|} = 29,58\,\%$$

The assessment results are summarized in Table 4.2 on page 71. ξ^{mx} is calculated using the equations from this subsection but with the following assumptions: $f_{ax} = 0\,kW$ and $\eta^{EL} = 100\,\%$, which results in $\eta^{mx}_\tau = 100\,\%$ and $\varepsilon^{mx} = \xi^{mx} = 51,51\,\%$.

A11.3 Absorption cooling using waste heat

The third cooling system considered in the exemplary comparison is an absorption cooling system, in which the desorption heat is being provided by waste heat, which is distributed by a district heating

Appendices

network. Figures A.11 and A.12 show the flow charts for the considered supply system

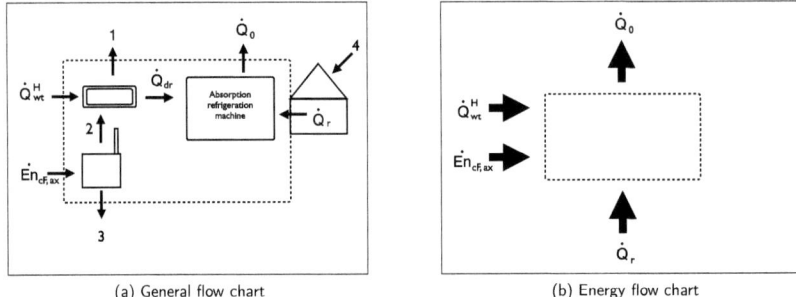

(a) General flow chart (b) Energy flow chart

Figure A.11: Flow charts of a cold supply system based on an absorption refrigeration machine - part 1

Figure A.11 (a) shows the different flows directly and indirectly relevant for the system evaluation.

Flow 1 designates heat losses from the district heating network, while flow 2 is the auxiliary power required to operate the district heating net. Flow 3 stands for losses from the power plant, while flow 4 equals the heat influx into the supply target, which has to be compensated for. Table A.13 summarizes the specific assumptions for the considered supply system.

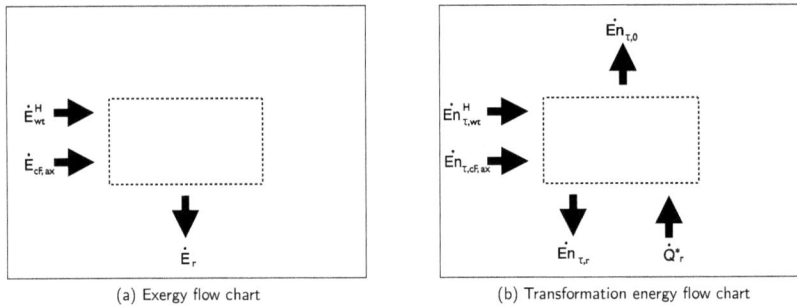

(a) Exergy flow chart (b) Transformation energy flow chart

Figure A.12: Flow charts of a cold supply system based on an absorption refrigeration machine - part 2

Appendices

Table A.13: Specific assumptions for the evaluation of a cooling supply system based on an absorption refrigeration machine operated with waste heat

Data	Symbol	Value	Units	Comment
Average temperature of heat flow absorbed from the supply target	$T_{a,l}$	$282,15$	K	9 °C as a mean temperature between the inflow at 6 °C and the exit flow at approximately 12 °C
Average temperature of heat flow discharged from the condenser & temperature of absorber heat output	$T_{a,h}$	$325,00$	K	
Auxiliary power for pump and equipment of the district heating net as a share of the waste heat input	f_{ax}	2%		
Driving heat ratio	$\dfrac{\dot{Q}_{dr}^{H,id}}{\dot{Q}_{dr}^{H}}$	$0,72$		The driving heat ratio has been chosen so that the *cold to heat ratio* $\dfrac{\dot{Q}_r}{\dot{Q}_{dr}}$ is approximately 0,5
Transport efficiency of district heating	η_{tr}	90%		The transport efficiency is a measure for $\dfrac{\dot{Q}_{dr}}{\dot{Q}_{wt}}$
Average driving heat temperature 90 °C	$T_{a,dr}$	$363,15\,K$	K	Heat influx at approximately 95 °C return flow temperature 85 °C

The driving heat flow is calculated directly from the cold to heat ratio as $\dot{Q}_{dr}^{H} = 10\,kW$ leading to a total waste heat input $\dot{Q}_{wt}^{H}$ into the system of:

$$\dot{Q}_{wt}^{H} = \frac{\dot{Q}_{dr}^{H}}{\eta_{tr}} = 11,11\,kW$$

The transformability of the driving heat $\tau_{dr} = 14,64\%$ and the compensation heat flow $\dot{Q}_{r}^{*} = 10,25\,kW$ are calculated according to the appropriate equations for heat flows below reference temperature given in Table A.2 on page 130. The combustible fuel input for the auxiliary power $\dot{E}n_{\tau,cF,ax}$ is obtained by solving:

$$\dot{E}n_{\tau,cF,ax} = \dot{Q}_{wt}^{H} \cdot \frac{f_{W,ax}}{\eta^{EL}} = 0,54\,kW$$

To calculate the effective compensation heat, the ideally required driving heat input has to be calculated. Since an absorption refrigeration machine can be considered a combination of a heat engine process and a compression refrigeration machine process (AHRAE, 1997), the ideally required driving heat is a function of the ideal coefficient of performance for a refrigeration machine operating between $T_{a,l}$ and $T_{a,h}$ and the maximum efficiency of a heat engine operating between T_{dr} and T_0. Figure A.13 shows the principal flow chart for this model of the absorption refrigeration machine. $\dot{Q}_{dr}^H$ stands for the driving heat input, while $\dot{Q}_r$ symbolizes the required heat transfer to the supply target. Flow 1 is the heat discharge from the heat engine, while flow 2 symbolizes the power transfer to the compression refrigeration machine. Flow 3 is the heat discharge from the compression refrigeration machine.

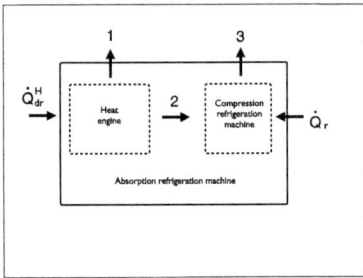

Figure A.13: Model of an absorption refrigeration machine as a combination of a heat engine and a compression refrigeration machine

Thus, the ideal heat to cold ratio can be calculated as a function of the ideal coefficient of performance of a compression refrigeration machine:

$$COP^{id} = \frac{T_{a,l}}{T_{a,h} - T_{a,l}} = 6,58$$

and the ideal electrical efficiency η^{id} of a heat engine operating between the provided temperatures:

$$\eta^{id} = 1 - \frac{T_{a,h}}{T_{a,dr}} = 10,50\,\%$$

The ideal cold to heat ratio is thus calculated as[11]:

$$\frac{\dot{Q}_r}{\dot{Q}_{dr}^{H,id}} = COP^{id} \cdot \eta^{id} = \frac{\dot{Q}_r}{\dot{W}^{id}} \cdot \frac{\dot{W}^{id}}{\dot{Q}_{dr}^{H,id}} = 0,69$$

The total ideally required heat discharge is a sum of the discharged heat from the absorber and the condenser of the refrigeration machine. It can be calculated on the basis of the energy balance for the absorption refrigeration machine as:

$$\dot{Q}_0^{id} = \dot{Q}_r + \dot{Q}_{dr}^{H,id} = \dot{Q}_r \cdot (1 + \frac{1}{0,69}) = 12,23\,kW$$

---

[11] The absorption temperature is a function of the working fluid, thus it is independent of the reference temperature.

Appendices

This results in an effective compensation heat of:

$$\Delta \dot{Q}_i^* = \dot{Q}_i^* - \dot{Q}_0^{id} = -1,97\,kW$$

, which means that the effective compensation heat is an output and has to considered in the numerator of the transformation energy efficiency. $\dot{Q}_0^{id}$ is a part of the real heat output of the compression refrigeration machine, which is labeled $\dot{Q}_0$ in the flowcharts.

On the basis of the given and the calculated values, the transformation energy efficiency can be calculated as:

$$\eta_\tau = \frac{\left|\dot{E}n_{\tau,r}\right| + \left|\Delta \dot{Q}_i^*\right|}{\dot{E}n_{\tau,cF} + \dot{E}n_{\tau,wt}} = 62,03\,\%$$

The exergetic efficiency is accordingly calculated as:

$$\varepsilon = \frac{\left|\dot{E}_r\right|}{\dot{E}_{cF} + \dot{E}_{wt}} = 11,73\,\%$$

ξ^{mx} is calculated using the equations from this subsection but with the following assumptions: $\eta_{tr} = 100\,\%$, $\frac{\dot{Q}_r}{\dot{Q}_{dr}^H} = \frac{\dot{Q}_r}{\dot{Q}_{dr}^{id,H}} = 0,69$, $f_{ax} = 0\,kW$ and $\eta^{EL} = 100\,\%$, which results in $\eta_\tau^{mx} = 100\,\%$ and $\varepsilon^{mx} = \xi^{mx} = 24,03\,\%$.

The results of the evaluation can be found in Table 4.2.

A12 On the use of average reference temperature

The use of average reference temperatures does not lead to a different evaluation than the calculation of the average exergy associated with a flow if this exergy value were calculated using time specific exergy values with the matching time specific reference temperatures[12]. This can be demonstrated using a simple example in which the average exergy is calculated from the time specific exergy of eight daily temperature measurements and compared with the exergy associated with the considered flow at average reference temperature. A conductive heat flow of $1\,kW$ at $340\,K$ is evaluated in the following.

---

[12]This statement is only valid for steady-state heat demand and the use of the average temperature for the time of operation, e.g. the heating period.

Table A.14: Exemplary daily exergy values for a conductive heat flow of $1\,kW$ at a temperature of $340\,K$

Daily reference temperature - T_0 [K]	Daily exergy - $\dot{E}$ [kW]
300,000	0,1176
296,000	0,1294
298,000	0,1235
305,000	0,1029
300,000	0,1176
308,000	0,0941
298,000	0,1235
304,000	0,1059

The average exergy flow $\dot{E}_a$ can be calculated as a function of the exergy flows at the different considered reference temperatures $\dot{E}$:

$$\dot{E}_a = \frac{\sum \dot{E}}{8} = 0,1143\,kW$$

The average reference temperature $T_{0,\,a}$ can be calculated similarly:

$$T_{0,\,a} = \frac{\sum T_0}{8} = 301,125\,K$$

Calculating the average exergy flow associated with the conductive heat flow using this value:

$$\dot{E}_a = 1\,kW \cdot \left(1 - \frac{301,125}{340}\right) = 0,1143\,kW$$

As expected, the two average exergy values are equal. Therefore, it can be assumed that the use of the average exergy for the evaluation of a heat flow that is independent of reference temperature does not lead to an error. Since the consideration of the influence of reference temperature changes is only a minor part of this work a mathematical proof of this statement is left to future researchers.

In contrast to the influence of changes of reference temperature that allows the use of average temperatures for the calculation of average exergy values without generating error the use of average pressures for ideal gases would not provide the matching average exergy values. As a consequence the correct assessment of the average mechanical exergy associated with a flow with low pressure differences to the environment would require a different approach. However, since mechanical exergy is not central for heat supply systems the influence of changing reference pressures on mechanical exergy and transformation energy is not discussed further.

Appendices

A13 Equations for transformability analysis of some common components

To allow easier process analysis using the transformability analysis method, the equations used to analyze some common components are presented in this section. The presentation is general and does not include numerical values, as it is intended only to lay a first basis for the application of transformability analysis. Tables A.16 and A.17 present a collection of the definitions of average in- and output transformabilities for processes operating mainly on the basis of thermal and mechanical exergy transfers. A precise consideration of chemical exergy is frequently not necessary for thermal and mechanical processes. Thus, chemical exergy is considered to be identical with the higher heating value, since the deviation of chemical exergy from higher heating value is usually very small.

The equations applicable for the calculation of the effective thermal transformabilities that are used to calculate the average transformabilities have to be chosen according to the temperature levels of the flows in relation to the reference temperature. The equations are derived in section 2.5 and summarized in Table A.4.

The expression $\Delta \dot{E}n_\tau^M$ is used to summarize the total mechanical transformation energy decrease in the considered process as an effect of pressure losses[13]. As the transformability of mechanical transformation energy is always $100\,\%$, the following equation is valid $\Delta \dot{E}n_\tau^M = \Delta \dot{E}^M$.

For a process with two inflows (subscript i) and two exit flows (subscript e), the total mechanical transformation energy decrease can be calculated as a function of the mass flows $\dot{m}$ and specific mechanical transformation energy en_τ^M as:

$$\Delta \dot{E}n_\tau^M = \dot{m}_1 \cdot \left(en_{\tau,1,i}^M - en_{\tau,1,e}^M\right) + \dot{m}_2 \cdot \left(en_{\tau,2,i}^M - en_{\tau,2,e}^M\right)$$

The general assumption for the following equations is that the effective compensation heat $\Delta \dot{Q}_i^* < 0$, which is the case if only mechanical compensation heat flows are considered for processes at $T > T_0$. For processes where $\Delta \dot{Q}_i^* > 0$ the effective compensation heat needs to be added to the denominator of the equation used for calculation of $\tau_{a,F}$ instead of being subtracted from and due to its negative sign effectively added to the denominator of the equation used for calculation of $\tau_{a,P}$. $\Delta \dot{Q}_i^*$ is thus always considered in the relevant ratios either as an input if $\Delta \dot{Q}_i^* > 0$ or as an output if $\Delta \dot{Q}_i^* < 0$. A discussion of the definition of the effective compensation heat can be found in subsection 3.2.3 on page 54 ff., while the definitions of ideally required refrigeration machines for some processes such as refrigeration machines, compressors, expanders and heat engines are discussed in appendix A6 on page 133 ff.

---

[13]Transformation energy cannot be destroyed. A decrease in mechanical transformation energy of a mass flow is always accompanied by a decrease of the matching compensation heat flows.

Appendices

Table A.15: Assumptions for the analysis of some basic processes

Process	Assumptions
General Assumptions	if not stated otherwise $\Delta \dot{Q}_i^* < 0$
Heat exchanger	$T > T_0$
Boiler	$T_F = T_0 \; ; \; T > T_0$
Compression heat pump	$T_h > T_l > T_0$
Compression refrigeration machine	$T_h > T_0 > T_l \; ; \; \Delta \dot{Q}_i^* > 0$
Compression refrigeration machine	$T_0 > T_h > T_l$
Heat engine	$T > T_0$
Expander	$T_i = T_0 \; ; \; \Delta \dot{Q}_i^* > 0$
Compressor	$T_i = T_0 \; ; \; \Delta \dot{Q}_i^* > 0$

Usually it is easier to calculate the transformability ratio ξ as the ratio of exergetic efficiency ε and the transformation energy efficiency η_τ:

$$\xi = \frac{\varepsilon}{\eta_\tau}$$

than to calculate effective thermal transformabilities and average thermal transformabilities. The approach of calculating ξ using effective thermal transformabilities τ^H and average transformabilities τ_a allows a deeper understanding of the implications of ξ, as its connections to the transformability associated with the considered mass flows are better visible, therefore it appears to be the better approach for the purpose of explanation.

A14 Analysis of a vapor-compression cascade refrigeration machine

The following section covers the calculation of the results that are discussed in section 4.5 on page 84 ff. The process flow chart is presented in Figure 4.3. Fluid property data is obtained from the Microsoft Excel Add-In Refprop 8.0 (NIST, 2007). The following tables summarize the provided temperatures and pressures as well as the thermodynamic data obtained from Refprop. The temperatures for flow 4 and for flow 14 have been calculated assuming an isenthalpic throttling of flows 3 and 13.

Table A.16.: Expressions for average transformabilities of heat exchangers, boilers, heat pumps and refrigeration machines

Process	ΔQ $\frac{kJ}{s}$	$\tau_{a,F}$	$\tau_{a,P}$
Heat exchanger	$\dot{m}_F \cdot (q_{F,i}^* - q_{F,e}^*)$ $+\dot{m}_P \cdot (q_{P,i}^* - q_{P,e}^*)$	$\dfrac{\left\| \dot{m}_P \cdot \left(e_{P,e}^T - e_{P,i}^T\right) \right\| + \left\| \Delta \dot{E}n_\tau^M \right\|}{\left\| \dot{m}_F \cdot \left(en_{\tau,F,i}^T - en_{\tau,F,e}^T\right) \right\| + \left\| \Delta \dot{E}n_\tau^M \right\|}$	$\dfrac{\left\| \dot{m}_P \cdot \left(e_{P,e}^T - e_{P,i}^T\right) \right\|}{\left\| \dot{m}_P \cdot \left(en_{\tau,P,e}^T - en_{\tau,P,i}^T\right) \right\| + \left\| \Delta \dot{Q}_i^* \right\|}$
Boiler[a]	$\dot{m}_F \cdot q_{F,i}^*$ $+\dot{m}_P \cdot (q_{P,i}^* - q_{P,e}^*)$	$\dfrac{\left\| \dot{m}_P \cdot \left(en_{\tau,P,e}^T - en_{\tau,P,i}^T\right) \right\| + \left\| \Delta \dot{E}n_\tau^M \right\|}{\dot{m}_F \cdot en_{\tau,F,i}^{CH} + \left\| \Delta \dot{E}n_\tau^M \right\|}$ $= 100\%$	$\dfrac{\left\| \dot{m}_P \cdot \left(e_{P,e}^T - e_{P,i}^T\right) \right\|}{\left\| \dot{m}_P \cdot \left(en_{\tau,P,e}^T - en_{\tau,P,i}^T\right) \right\| + \left\| \Delta \dot{Q}_i^* \right\|}$
Compression heat pump	$\dot{m}_l \cdot (q_{l,i}^* - q_{l,e}^*)$ $+\dot{m}_h \cdot (q_{h,i}^* - q_{h,e}^*)$	$\dfrac{\left\| \dot{m}_l \cdot \left(e_{l,i}^T - e_{l,e}^T\right) \right\| + \left\| \Delta \dot{E}_\tau^M \right\| + \left\| \dot{W} \right\|}{\left\| \dot{m}_l \cdot \left(en_{\tau,l,i}^T - en_{\tau,l,e}^T\right) \right\| + \left\| \Delta \dot{E}n_\tau^M \right\| + \left\| \dot{W} \right\|}$	$\dfrac{\left\| \dot{m}_h \cdot \left(e_{h,e}^T - e_{h,i}^T\right) \right\|}{\left\| \dot{m}_h \cdot \left(en_{\tau,h,e}^T - en_{\tau,h,i}^T\right) \right\| + \left\| \Delta \dot{Q}_i^* \right\|}$ $= \tau_l^H$
Compression refrigeration machine $T_h > T_0 > T_l$	$\dot{m}_l \cdot (q_{l,i}^* - q_{l,e}^*)$ $+\dot{m}_h \cdot \left(en_{\tau,h,i}^{T,id} - en_{\tau,h,e}^{T,id}\right)$	$\dfrac{\left\| \dot{W} \right\| + \left\| \Delta \dot{E}^M \right\|}{\left\| \dot{W} \right\| + \left\| \Delta \dot{E}n_\tau^M \right\| + \left\| \Delta \dot{Q}_i^* \right\|}$	$\dfrac{\left\| \dot{m}_l \cdot \left(e_{l,e}^T - e_{l,i}^T\right) \right\|}{\left\| \dot{m}_l \cdot \left(en_{\tau,l,e}^T - en_{\tau,l,i}^T\right) \right\| + \left\| \Delta \dot{Q}_i^* \right\|}$
Compression refrigeration machine $T_0 > T_h > T_l$	$\dot{m}_l \cdot (q_{l,i}^* - q_{l,e}^*)$ $+\dot{m}_h \cdot (q_{h,i}^* - q_{h,e}^*)$	$\dfrac{\left\| \dot{m}_h \cdot \left(en_{\tau,h,i}^T - en_{\tau,h,e}^T\right) \right\| + \left\| \Delta \dot{E}^M \right\| + \left\| \dot{W} \right\|}{\left\| \dot{m}_h \cdot \left(en_{\tau,h,i}^T - en_{\tau,h,e}^T\right) \right\| + \left\| \Delta \dot{E}n_\tau^M \right\| + \left\| \dot{W} \right\|}$	$\dfrac{\left\| \dot{m}_l \cdot \left(e_{l,e}^T - e_{l,i}^T\right) \right\|}{\left\| \dot{m}_l \cdot \left(en_{\tau,l,e}^T - en_{\tau,l,i}^T\right) \right\| + \left\| \Delta \dot{Q}_i^* \right\|}$

[a] These equations are strictly valid only under the assumption that neither reactands nor products are associated with chemical exergy.

Table A.17: Expressions for average transformabilities of heat engines, expanders and compressors

Process	$\Delta \dot{Q}_i^*$ $\frac{kJ}{s}$	$\tau_{a,F}$	$\tau_{a,P}$
Heat engine	$\dot{m}_l \cdot (q_{l,i}^* - q_{l,e}^*)$ $+ \dot{m}_h \cdot (q_{h,i}^* - q_{h,e}^*)$ $+ \dot{m}_l \cdot \left(en_{\tau,l,i}^{T,id} - en_{\tau,l,e}^{T,id}\right)$	$\dfrac{\left|\dot{m}_h \cdot \left(en_{\tau,h,i}^T - en_{\tau,h,e}^T\right)\right| + \left|\Delta \dot{E}_{\tau}^M\right|}{\left|\dot{m}_h \cdot \left(en_{\tau,h,i}^T - en_{\tau,h,e}^T\right)\right| + \left|\Delta \dot{E}n_{\tau}^M\right|}$	
Expander	$\dot{m}_e \cdot (q_i^* - q_e^*) + \dot{Q}_i^{id}$	$\dfrac{\left|\dot{m}_e \cdot (e_i^M - e_e^M)\right|}{\left|\dot{m}_e \cdot \left(en_{\tau,i}^M - en_{\tau,e}^M\right)\right| + \left|\Delta \dot{Q}_i^*\right|}$	$\dfrac{|\dot{W}|}{|\dot{W}| + |\Delta \dot{Q}_i^*|}$
Compressor	$\dot{m}_e \cdot (q_i^* - q_e^*) + \dot{Q}_i^{id}$	$\dfrac{|\dot{W}|}{|\dot{W}| + |\Delta \dot{Q}_i^*|}$	$\dfrac{\left|\dot{m}_e \cdot \left(e_e^T - e_i^T\right)\right| + |\dot{W}|}{\dot{m}_e \cdot \left(\left(e_e^M - e_i^M\right| + \left|e_e^T - e_i^T\right|\right) + \left|en_{\tau,e}^T - en_{\tau,i}^T\right|\right)}$

Table A.18: Equations for the analysis of heat exchangers, boilers, heat pumps and refrigeration machines

Process	η_T	ξ	ε
Heat exchanger	$\dfrac{\lvert \dot{m}_P \cdot (en^T_{T,P,e} - en^T_{T,P,i}) \rvert + \lvert \Delta \dot{Q}^*_i \rvert}{\lvert \dot{m}_F \cdot (en^T_{T,F,i} - en^T_{T,F,e}) \rvert + \lvert \Delta \dot{E} n^M_T \rvert}$	$\dfrac{T_{a,P}}{T_{a,F}}$	$\dfrac{\lvert \dot{m}_P \cdot (e^T_{P,e} - e^T_{P,i}) \rvert}{\lvert \dot{m}_F \cdot (e^T_{F,i} - e^T_{F,e}) \rvert + \lvert \Delta \dot{E}^M \rvert}$
Boiler	$\dfrac{\lvert \dot{m}_P \cdot (en^T_{T,P,e} - en^T_{T,P,i}) \rvert + \lvert \Delta \dot{Q}^*_i \rvert}{\lvert \dot{m}_F \cdot en^{CH}_{T,F} \rvert}$	$\dfrac{T_{a,P}}{T_{a,F}}$	$\dfrac{\lvert \dot{m}_P \cdot (e^T_{P,e} - e^T_{P,i}) \rvert}{\lvert \dot{m}_F \cdot e^{CH}_{P,i} \rvert + \lvert \Delta \dot{E}^M \rvert}$
Compression heat pump	$\dfrac{\lvert \dot{m}_h \cdot (en^T_{T,h,e} - en^T_{T,h,i}) \rvert - \lvert \Delta \dot{Q}^*_i \rvert}{\lvert \dot{m}_l \cdot (en^T_{T,l,i} - en^T_{T,l,e}) \rvert + \lvert \Delta \dot{E} n^M_T \rvert + \lvert \dot{W} \rvert}$	$\dfrac{T_{a,P}}{T_{a,F}}$	$\dfrac{\lvert \dot{m}_h \cdot (e^T_{h,e} - e^T_{h,i}) \rvert}{\lvert \dot{m}_l \cdot (e^T_{l,i} - e^T_{l,e}) \rvert + \lvert \Delta \dot{E}^M \rvert + \lvert \dot{W} \rvert}$
Compression refrigeration machine $T_h > T_0 > T_l$	$\dfrac{\lvert \dot{m}_l \cdot (en^T_{T,l,e} - en^T_{T,l,i}) \rvert + \lvert \Delta \dot{Q}^*_i \rvert}{\lvert \dot{W} \rvert + \lvert \Delta \dot{E} n^M_T \rvert}$	$\dfrac{T_{a,P}}{T_{a,F}}$	$\dfrac{\lvert \dot{m}_l \cdot (e^T_{l,i} - e^T_{l,e}) \rvert + \lvert \Delta \dot{E}^M \rvert}{\lvert \dot{W} \rvert + \lvert \Delta \dot{E}^M \rvert}$
Compression refrigeration machine $T_0 > T_h > T_l$	$\dfrac{\lvert \dot{m}_l \cdot (en^T_{T,l,e} - en^T_{T,l,i}) \rvert + \lvert \Delta \dot{Q}^*_i \rvert}{\lvert \dot{m}_U \cdot (en^T_{T,h,i} - en^T_{T,h,e}) \rvert + \lvert \Delta \dot{E} n^M_T \rvert + \lvert \dot{W} \rvert}$	$\dfrac{T_{a,P}}{T_{a,F}}$	$\dfrac{\lvert \dot{m}_l \cdot (e^T_{l,e} - e^T_{l,i}) \rvert}{\lvert \dot{m}_U \cdot (e^T_{h,i} - e^T_{h,e}) \rvert + \lvert \Delta \dot{E}^M \rvert + \lvert \dot{W} \rvert}$

Table A.19: Equations for the analysis of heat engines, expanders and compressors

Process	η_T	ζ	ε
Heat engine	$\dfrac{\|\dot{W}\| + \|\Delta \dot{Q}_i^*\|}{\|\dot{m}_h \cdot \left(en_{\tau,h,i}^T - en_{\tau,h,e}^T\right)\| + \|\Delta \dot{En}_\tau^M\|}$	$\dfrac{T_{a,e}}{T_{a,i}}$	$\dfrac{\|\dot{W}\|}{\|\dot{m}_h \cdot \left(e_{h,i}^T - e_{h,e}^T\right)\| + \|\Delta \dot{En}_\tau^M\|}$
Expander	$\dfrac{\|\dot{m}_e \cdot \left(en_{\tau,e}^T - en_{\tau,i}^T\right)\| + \|\dot{W}\|}{\|\dot{m}_e \cdot \left(en_{\tau,e}^M - en_{\tau,i}^M\right)\| + \|\Delta \dot{Q}_i^*\|}$	$\dfrac{T_{a,e}}{T_{a,i}}$	$\dfrac{\|\dot{m}_e \cdot \left(e_e^T - e_i^T\right)\| + \|\dot{W}\|}{\|\dot{m}_e \cdot \left(e_e^M - e_i^M\right)\|}$
Compressor	$\dfrac{\|\dot{m}_e \cdot \left(\left(en_{\tau,e}^M - en_{\tau,i}^M\right) + \|en_{\tau,e}^T - en_{\tau,i}^T\|\right)\|}{\|\dot{W}\| + \|\Delta \dot{Q}_i^*\|}$	$\dfrac{T_{a,e}}{T_{a,i}}$	$\dfrac{\|\dot{m}_e \cdot \left(\|e_e^M - e_i^M\| + \|e_e^T - e_i^T\|\right)\|}{\|\dot{W}\|}$

Appendices

Table A.20: Thermodynamic data of air

Flow	Substance	$\dot{m}$	t	p	T	p	h	h_{T0}	s	s_{T0}
		$\frac{kg}{s}$	°C	bar	K	MPa	$\frac{kJ}{kg}$	$\frac{kJ}{kg}$	$\frac{kJ}{kg \cdot K}$	$\frac{kJ}{kg \cdot K}$
A	air	0,50	25,00	2,00	298,15	0,20	298,22	298,22	6,66	6,66
B	air	0,50	-40,00	1,75	233,15	0,18	232,78	298,28	6,46	6,70
C	air	0,50	-60,00	1,50	213,15	0,15	212,72	298,34	6,41	6,75
D	air	0,50	5,00	1,25	278,15	0,13	278,26	298,39	6,73	6,80

Table A.21: Thermodynamic data of ethane (R170), the working fluid of the low cascade

Flow	Substance	$\dot{m}$	t	p	T	p	h	h_{T0}	s	s_{T0}
		$\frac{kg}{s}$	°C	bar	K	MPa	$\frac{kJ}{kg}$	$\frac{kJ}{kg}$	$\frac{kJ}{kg \cdot K}$	$\frac{kJ}{kg \cdot K}$
1	ethane	0,03	-65,00	2,59	208,15	0,26	516,81	664,73	2,54	3,13
2	ethane	0,03	51,00	14,60	324,15	1,46	690,99	638,89	2,76	2,59
3	ethane	0,03	-25,00	14,60	248,15	1,46	168,66	638,89	0,77	2,59
4	ethane	0,03	-69,18	2,59	203,97	0,26	168,66	664,73	0,84	3,13

Table A.22: Thermodynamic data of propane (R290), the working fluid of the high cascade

Flow	Substance	$\dot{m}$	t	p	T	p	h	h_{T0}	s	s_{T0}
		$\frac{kg}{s}$	°C	bar	K	MPa	$\frac{kJ}{kg}$	$\frac{kJ}{kg}$	$\frac{kJ}{kg \cdot K}$	$\frac{kJ}{kg \cdot K}$
11	propane	0,05	-25,00	1,64	248,15	0,16	548,17	628,51	2,46	2,75
12	propane	0,05	63,00	10,84	336,15	1,08	673,04	265,11	2,56	1,22
13	propane	0,05	25,00	10,84	298,15	1,08	265,11	265,11	1,22	1,22
14	propane	0,05	-30,60	1,64	242,55	0,16	265,11	628,51	1,29	2,75

To obtain the mass flows of ethane in the low cascade (subscript LC), the given refrigeration capacity of $\dot{Q}_l = 10,00\,kW$ is used:

$$\dot{m}_{LC} = \frac{\dot{Q}_l}{h_4 - h_1}$$

The mass flow of propane in the high cascade (subscript HC) is calculated in a second step using the energy balance of the condenser/evaporator (CD-EV), which leads to the following equation:

$$\dot{m}_{HC} = \dot{m}_{LC} \cdot \frac{h_3 - h_2}{h_{14} - h_{11}}$$

The mass flows of air have been calculated based on the refrigeration capacity as:

Appendices

$$\dot{m}_{air} = \frac{\dot{Q}_l}{(h_B - h_C)}$$

Using the data provided in Tables A.20, A.21 and A.22 as well as the equations for transformation energy and compensation heat flows presented in Table A.2 on page 130 and exergy equations discussed in chapter 1 the following data is obtained for exergies, transformation energies and compensation heat flows associated with the mass flows under consideration:

Table A.23: Exergy, transformation energy and compensation heat flows associated with air flows in the vapor-cascade refrigeration machine

Flow	Substance	e	e^T	e^M	en_τ	en_τ^T	en_τ^M	$q^{*,T}$	$q^{*,M}$
		$\frac{kJ}{kg}$	$\frac{kJ}{kg}$	$\frac{kJ}{kg}$	$\frac{kJ}{kg}$	$\frac{kJ}{kg}$	$\frac{kJ}{kg}$	$\frac{kJ}{kg}$	$\frac{kJ}{kg}$
A	air	59,31	0,00	59,31	59,31	0,00	59,31	0,00	-59,54
B	air	56,27	8,38	47,89	121,76	73,88	47,89	-139,37	-48,06
C	air	49,87	15,17	34,70	135,48	100,79	34,70	-186,40	-34,81
D	air	19,80	0,71	19,10	39,94	-20,13	19,10	0,00	-19,15

Table A.24: Exergy, transformation energy and compensation heat flows associated with ethane flows in the vapor-cascade refrigeration machine

Flow	Substance	e	e^T	e^M	en_τ	en_τ^T	en_τ^M	$q^{*,T}$	$q^{*,M}$
		$\frac{kJ}{kg}$	$\frac{kJ}{kg}$	$\frac{kJ}{kg}$	$\frac{kJ}{kg}$	$\frac{kJ}{kg}$	$\frac{kJ}{kg}$	$\frac{kJ}{kg}$	$\frac{kJ}{kg}$
1	ethane	104,93	27,46	77,47	252,85	175,38	77,47	-323,30	-80,55
2	ethane	214,41	2,15	212,26	264,36	52,10	212,26	0,00	-241,18
3	ethane	285,89	73,63	212,26	756,11	543,86	212,26	-1014,08	-241,18
4	ethane	265,59	188,12	77,47	761,66	684,19	77,47	-1180,26	-80,55

Table A.25: Exergy, transformation energy and compensation heat flows associated with propane flows in the vapor-cascade refrigeration machine

Flow	Substance	e	e^T	e^M	en_τ	en_τ^T	en_τ^M	$q^{*,T}$	$q^{*,M}$
		$\frac{kJ}{kg}$	$\frac{kJ}{kg}$	$\frac{kJ}{kg}$	$\frac{kJ}{kg}$	$\frac{kJ}{kg}$	$\frac{kJ}{kg}$	$\frac{kJ}{kg}$	$\frac{kJ}{kg}$
11	propane	34,65	7,45	27,20	115,00	87,80	27,20	-168,14	-29,06
12	propane	128,92	10,20	118,72	526,65	407,93	118,72	0,00	-483,98
13	propane	118,72	0,00	118,72	118,72	0,00	118,72	0,00	-483,98
14	propane	99,42	72,22	27,20	462,83	435,63	27,20	-799,03	-29,06

Appendices

The work flow input into the low cascade compressor $\dot{W}_{LC}$ can be calculated using the given isentropic efficiency of 63 % as:

$$\dot{W}_{LC} = \frac{\dot{m}_2 \cdot (h_2 - h_1)}{63\%} = 7,94\,kW$$

Similarly, the work flow input into the high cascade compressor P_{HC} can be calculated using the given isentropic efficiency of 73 % as:

$$\dot{W}_{HC} = \frac{\dot{m}_{12} \cdot (h_{12} - h_{11})}{73\%} = 9,07\,kW$$

The specific compensation heat q^* of a flow is obtained as:

$$q^* = q^{*,T} + q^{*,M}$$

Using data from Tables A.23, A.24 and A.25 and the equations summarized in Tables A.27 and A.28 the results collected in Tables 4.6 on page 86 and 4.7 on page 87 have been calculated. The basis on which the equations in Tables A.27 and A.28 have been obtained are the equations presented in Tables A.2 on page 130, A.3 on page 131 and A.4 on page 132.

Table A.26 summarizes the equations used for the calculation of the ideally required heat flows. These heat flows are necessary for process operation and have been calculated according to the discussion in subsection 3.2.3 on page 54 ff. and the following subsections.

Table A.26: Ideally required heat flows

Process	$\dot{Q}^{id}_{0,i}$
	$\frac{kJ}{s}$
Low cascade - evaporator [EV] ($4 \to 1/B \to C$)	-
Low cascade - compressor [LCCM] ($1 \to 2$)	$\dot{m}_1 \cdot (h_1 - h_2 + e_2 - e_1)$
Intercascade condenser/evaporator [CD-EV] ($2 \to 3/14 \to 11$)	-
High cascade - compressor [HCCM] ($11 \to 12$)	$\dot{m}_{11} \cdot (h_{11} - h_{12} + e_{12} - e_{11})$
Low cascade - throttle [TV1] ($3 \to 4$)	-
Low cascade - throttle [TV2] ($13 \to 14$)	-
Total process	$\dot{m}_B \cdot (h_B - h_C) \cdot \left(1 + \frac{T_{a,h} - T_{a,l}}{T_{a,l}}\right)$

Appendices

A15 Basic data for the calculation of the exemplary ExergyFingerprints

Table A.30 shows the basic data which has been used for the calculation of the average transformability values and the transformation energies for presentation in in the ExergyFingerprints discussed in subsection 4.6 on page 88 ff. A detailed calculation of the values is not presented, since it can be performed in analogy to the calculations of the previously discussed examples in appendix A10 on page 143 ff.

Table A.30: Basic data for the calculation of the exemplary ExergyFingerprint demand structure

Type of data		Values		Source
Application of terminal energy provided to german Households in percent of the average terminal energy demand.	Space heating	74%		BMWI 2008
	Warm water	12%		
	Process heat	5%		
	Mechanical and electrical energy	8%		
	Lighting	2%		
Total terminal energy consumption in Germany (2006)		2.660,00	$\frac{PJ}{a}$	
Inhabitants in Germany 2006		82.314.900		DeStatis 2009
Average annual terminal energy demand for domestic use of a German citizen		8,97	$\frac{MWh}{a}$	$\frac{2600 \frac{PJ}{a}}{8.314.900}$

The basis for calculating the effective compensation heat flow are the specific compensation heats $\dot{q}^*$ associated with the considered massflows. They are calculated as the sum of the specific thermal compensation heat flows $q^{*,T}$ and the specific mechanical compensation heat flow $\dot{q}^{*,M}$ associated with the considered mass flow.

Table A.27: Effective thermal transformabilities and effective compensation heat flows - equations

Process	$\Delta \dot{Q}_i^*$ $\left[\frac{kJ}{s}\right]$	τ_i^H	τ_e^H
Low cascade - evaporator [EV] ($4 \to 1/B \to C$)	$\dot{m}_C \cdot (q_B^* - q_C^*)$ $+ \dot{m}_1 \cdot (q_4^* - q_1^*)$	$1 - \dfrac{(h_4 - h_{T0,4}) - (h_1 - h_{T0,1})}{T_0 \cdot [(s_4 - s_{T0,4}) - (s_1 - s_{T0,1})]}$	$1 - \dfrac{(h_B - h_{T0,B}) - (h_C - h_{T0,C})}{T_0 \cdot [(s_B - s_{T0,B}) - (s_C - s_{T0,C})]}$
Low cascade - compressor [LCCM] ($1 \to 2$)	$\dot{m}_1 \cdot (q_1^* - q_2^*) + \dot{Q}_{0,i}^{id}$	$1 - \dfrac{(h_1 - h_{T0,1}) + T_0 \cdot (s_2 - s_{T0,2})}{(h_2 - h_{T0,2}) + T_0 \cdot (s_1 - s_{T0,1})}$	-
Intercascade condenser/evaporator [CD-EV] ($2 \to 3 / 14 \to 11$)	$\dot{m}_3 \cdot (q_2^* - q_3^*)$ $+ \dot{m}_{11} \cdot (q_{14}^* - q_{11}^*)$	$1 - \dfrac{(h_{14} - h_{T0,14}) - (h_{11} - h_{T0,11})}{T_0 \cdot [(s_{14} - s_{T0,14}) - (s_{11} - s_{T0,11})]}$	$1 - \dfrac{(h_2 - h_{T0,2}) + T_0 \cdot (s_3 - s_{T0,3})}{(h_3 - h_{T0,3}) + T_0 \cdot (s_2 - s_{T0,2})}$
High cascade - compressor [HCCM] ($11 \to 12$)	$\dot{m}_{11} \cdot (q_{11}^* - q_{12}^*) + \dot{Q}_{0,i}^{id}$	$1 - \dfrac{(h_{11} - h_{T0,11}) + T_0 \cdot (s_{12} - s_{T0,12})}{(h_{12} - h_{T0,12}) + T_0 \cdot (s_{11} - s_{T0,11})}$	
Low cascade - throttle [TV1] ($3 \to 4$)	$\dot{m}_3 \cdot (q_3^* - q_4^*)$	-	$1 - \dfrac{(h_3 - h_{T0,3}) - (h_4 - h_{T0,4})}{T_0 \cdot [(s_3 - s_{T0,3}) - (s_4 - s_{T0,4})]}$
High cascade - throttle [TV2] ($13 \to 14$)	$\dot{m}_{13} \cdot (q_{13}^* - q_{14}^*)$	-	$1 - \dfrac{(h_{14} - h_{T0,14}) + T_0 \cdot (s_{13} - s_{T0,13})}{(h_{13} - h_{T0,13}) + T_0 \cdot (s_{14} - s_{T0,14})}$
Total process	$\dot{m}_C \cdot (q_B^* - q_C^*) + \dot{Q}_{0,i}^{id}$	-	$1 - \dfrac{(h_B - h_{T0,B}) - (h_C - h_{T0,C})}{T_0 \cdot [(s_B - s_{T0,B}) - (s_C - s_{T0,C})]}$

Based on the flow chart in Figure 4.3 and the results presented in Tables A.23, A.24 and A.25 the transformation energy efficiency, the transformability ratio and the exergetic efficiency in Table 4.6 on page 86 are obtained. The equations which have been used to calculate these results are summarized in Table A.29.

The exergy associated with the heat flow discharged to water in the condenser of the high cascade is considered to be fully destroyed, thus the heat discharge of the total process can be considered a heat transfer at reference temperature. The relevant ideally required heat flows are summarized in Table A.27.

Table A.28: Average in- and output transformabilities - equations

Process	$\tau_{a,F}$	$\tau_{a,P}$
Low cascade - evaporator [EV] $(4 \to 1/B \to C)$	$\dfrac{\dot{m}_1 \cdot (e_4^T - e_1^T) + \dot{m}_C \cdot (e_B^M - e_C^M)}{\dot{m}_1 \cdot (en_{\tau,4}^T - en_{\tau,1}^T) + \dot{m}_C \cdot (en_{\tau,B}^M - en_{\tau,C}^M)}$	$\dfrac{\dot{m}_C \cdot (e_C^T - e_B^T)}{\dot{m}_C \cdot (en_{\tau,C}^T - en_{\tau,B}^T) + \left|\Delta \dot{Q}_i^*\right|}$
Low cascade - compressor [LCCM] $(1 \to 2)$	$\dfrac{\dot{m}_1 \cdot (e_1^T - e_2^T) + \dot{W}_{LC}}{\dot{m}_1 \cdot (en_{\tau,1}^T - en_{\tau,2}^T) + \dot{W}_{LC}}$	$\dfrac{\dot{m}_2 \cdot (e_2^T - e_1^T)}{\dot{m}_1 \cdot (en_{\tau,2}^T - en_{\tau,1}^T) + \left|\Delta \dot{Q}_i^*\right|}$
Intercascade condenser/evaporator [CD-EV] $(2 \to 3/14 \to 11)$	τ_i^H	$\dfrac{\dot{m}_3 \cdot (e_3^T - e_2^T)}{\dot{m}_3 \cdot (en_{\tau,3}^T - en_{\tau,2}^T) + \left|\Delta \dot{Q}_i^*\right|}$
High cascade - compressor [HCCM] $(11 \to 12)$	$\dfrac{\dot{W}_{HC}}{\dot{W}_{HC} + \Delta \dot{Q}_i^*}$	$\dfrac{(e_{12}^T - e_{11}^T) + (e_{12}^M - e_{11}^M)}{\dot{m}_3 \cdot (en_{\tau,3}^T - en_{\tau,2}^T) + (en_{\tau,12}^M - en_{\tau,11}^M)}$
Low cascade - throttle [TV1] $(3 \to 4)$	$\dfrac{\dot{m}_3 \cdot (e_3^M - e_4^M)}{\dot{m}_3 \cdot (en_{\tau,3}^M - en_{\tau,4}^M) + \Delta \dot{Q}_i^*}$	τ_e^H
Low cascade - throttle [TV2] $(13 \to 14)$	$\dfrac{\dot{m}_{13} \cdot (e_{13}^M - e_{14}^M)}{\dot{m}_3 \cdot (en_{\tau,13}^M - en_{\tau,14}^M) + \Delta \dot{Q}_i^*}$	τ_e^H
Total process	$\dfrac{\dot{W}_{HC} + \dot{W}_{LC} + \dot{m}_C \cdot (e_B^M - e_C^M)}{\dot{W}_{HC} + \dot{W}_{LC} + \dot{m}_C \cdot (en_{\tau,B}^M - en_{\tau,C}^M) + \Delta \dot{Q}_i^*}$	$\dfrac{\dot{m}_C \cdot (e_C^T - e_B^T)}{\dot{m}_C \cdot (en_{\tau,C}^T - en_{\tau,B}^T)}$

Table A.29: Equations for the evaluation of a vapor-compression cascade refrigeration machine

Process	η_τ	ζ	ε
Low cascade - evaporator [EV] ($4 \rightarrow 1/B \rightarrow C$)	$\dfrac{\dot{m}_B \cdot \left(en_{\tau,C}^T - en_{\tau,B}^T\right)}{\dot{m}_1 \cdot \left(en_{\tau,4}^T - en_{\tau,1}^T\right) + \dot{m}_C \cdot \left(en_{\tau,B}^M - en_{\tau,C}^M\right)}$	$\dfrac{T_a, P}{T_a, F}$	$\dfrac{\dot{m}_B \cdot \left(e_C^T - e_B^T\right)}{\dot{m}_1 \cdot \left(e_4^T - e_1^T\right) + \dot{m}_C \cdot \left(e_B^M - e_C^M\right)}$
Low cascade - compressor [LCCM] ($1 \rightarrow 2$)	$\dfrac{\dot{m}_2 \cdot \left(en_{\tau,2}^M - en_{\tau,1}^M\right) + \dot{m}_C \cdot \left(en_{\tau,B}^M - en_{\tau,C}^M\right)}{\dot{m}_2 \cdot \left(en_{\tau,1}^T - en_{\tau,2}^T\right) + \left|\Delta \dot{Q}_i^*\right|}$	$\dfrac{T_a, P}{T_a, F}$	$\dfrac{\dot{m}_2 \cdot \left(e_2^M - e_1^M\right)}{\dot{m}_2 \cdot \left(e_1^T - e_2^T\right) + \dot{W}_{LC}}$
Intercascade condenser/evaporator [CD-EV] ($2 \rightarrow 3/14 \rightarrow 11$)	$\dfrac{\dot{m}_3 \cdot \left(en_{\tau,3}^T - en_{\tau,2}^T\right) + \left|\Delta \dot{Q}_i^*\right|}{\dot{m}_{11} \cdot \left(en_{\tau,14}^T - en_{\tau,11}^T\right)}$	$\dfrac{T_a, P}{T_a, F}$	$\dfrac{\dot{m}_3 \cdot \left(e_3^T - e_2^T\right)}{\dot{m}_{11} \cdot \left(e_{14}^T - e_{11}^T\right)}$
High cascade - compressor ($11 \rightarrow 12$) [HCCM]	$\dfrac{\dot{m}_2 \cdot \left(en_{\tau,2}^M - en_{\tau,1}^M + en_{\tau,2}^T - en_{\tau,1}^T\right)}{\dot{W}_{HC} + \Delta \dot{Q}_i^*}$	$\dfrac{T_a, P}{T_a, F}$	$\dfrac{\dot{m}_2 \cdot \left(e_2^M - e_1^M + e_2^T - e_1^T\right)}{\dot{W}_{HC}}$
Low cascade - throttle [TV1] ($3 \rightarrow 4$)	$\dfrac{\dot{m}_3 \cdot \left(en_{\tau,4}^T - en_{\tau,3}^T\right)}{\dot{m}_3 \cdot \left(en_{\tau,3}^M - en_{\tau,4}^M\right) + \Delta \dot{Q}_i^*}$	$\dfrac{T_a, P}{T_a, F}$	$\dfrac{\dot{m}_3 \cdot \left(e_4^T - e_3^T\right)}{\dot{m}_3 \cdot \left(e_3^M - e_4^M\right)}$
Low cascade - throttle [TV2] ($13 \rightarrow 14$)	$\dfrac{\dot{m}_{13} \cdot \left(en_{\tau,14}^T - en_{\tau,13}^T\right)}{\dot{m}_{13} \cdot \left(en_{\tau,13}^M - en_{\tau,14}^M\right) + \Delta \dot{Q}_i^*}$	$\dfrac{T_a, P}{T_a, F}$	$\dfrac{\dot{m}_{13} \cdot \left(e_{14}^T - e_{13}^T\right)}{\dot{m}_{13} \cdot \left(e_{13}^M - e_{14}^M\right)}$
Total process	$\dfrac{\dot{m}_B \cdot \left(en_{\tau,C}^T - en_{\tau,B}^T\right)}{\dot{m}_C \cdot \left(en_{\tau,B}^M - en_{\tau,C}^M\right) + \dot{W}_{LC} + \dot{W}_{HC} + \Delta \dot{Q}_i^*}$	$\dfrac{T_a, P}{T_a, F}$	$\dfrac{\dot{m}_B \cdot \left(e_C^T - e_B^T\right)}{\dot{m}_C \cdot \left(e_B^M - e_C^M\right) + \dot{W}_{LC} + \dot{W}_{HC}}$

Appendices

Table A.31: Required temperature levels and transformabilities as basic data for the exemplary ExergyFingerprints

Type of useful energy	Required average temperature of useful energy [°C]	Average temperature [K]	Average transformability
Electrical power	-	-	100%
Hot tap water	45,00	318	13,52%
Room heat	20,00	293	6,14%
Reference temperature	0	275	0,00%
Average temperature of process heat (Cooking, Washing, Drying, Dishwashing)	135,00	408,15	32,62%

Process-specific data that has been used for the calculation of the ExergyFingerprints is presented in appendices A10.1 on page 144 ff. and A10.4 on page 149 ff.

VDM Verlagsservicegesellschaft mbH

Die VDM Verlagsservicegesellschaft sucht für wissenschaftliche Verlage abgeschlossene und herausragende

Dissertationen, Habilitationen, Diplomarbeiten, Master Theses, Magisterarbeiten usw.

für die kostenlose Publikation als Fachbuch.

Sie verfügen über eine Arbeit, die hohen inhaltlichen und formalen Ansprüchen genügt, und haben Interesse an einer honorarvergüteten Publikation?

Dann senden Sie bitte erste Informationen über sich und Ihre Arbeit per Email an *info@vdm-vsg.de*.

Sie erhalten kurzfristig unser Feedback!

VDM Verlagsservicegesellschaft mbH
Dudweiler Landstr. 99   Telefon +49 681 3720 174
D - 66123 Saarbrücken   Fax   +49 681 3720 1749
www.vdm-vsg.de

Die VDM Verlagsservicegesellschaft mbH vertritt

Printed by Books on Demand GmbH, Norderstedt / Germany